S0-BKW-867

Nonrelativistic Mechanics

Modern Physics Monograph Series
FELIX M. H. VILLARS, EDITOR

ROBERT J. FINKELSTEIN
Nonrelativistic Mechanics, 1973

ROBERT T. SCHUMACHER
Introduction to Magnetic Resonance: Principles and Applications, 1970

In Preparation:

HELLMUT J. JURETSCHKE
Crystal Physics

Nonrelativistic Mechanics

Robert J. Finkelstein
University of California, Los Angeles

1973
W. A. BENJAMIN, INC.
ADVANCED BOOK PROGRAM
Reading, Massachusetts

London · Amsterdam · Don Mills, Ontario · Sydney · Tokyo

Library of Congress Cataloging in Publication Data

Finkelstein, Robert J 1916–
Nonrelativistic mechanics.

(Modern physics monograph series)
Includes bibliographies.
1. Quantum theory. 2. Mechanics. I. Title.
QC174.1.F5 530.1'2 72-5557

ISBN 0–805–32551–9
ISBN 0–805–32552–7 (pbk.)

Reproduced by W. A. Benjamin, Inc., Advanced Book Program, Reading, Massachusetts, from camera-ready copy prepared by the author.

Manufactured in the United States of America

HA 2/73 32551

CONTENTS

EDITOR'S FOREWORD

Education in physics is going through a phase of rapid evolution. On the frontier of the field new information is literally pouring in, new perspectives are opening up, and new concepts are emerging. For the student, the distance to be covered from freshman year to graduate research work appears to be ever expanding.

Professional education in physics therefore must deal with the very real problem of the need for thoughtful condensation of the material presented, and with the question of what may and should reasonably be achieved in the years between the introductory and the research level.

It is generally agreed, on the one hand, that a thorough presentation of the fundamentals of both classical and quantum physics is essential. On the other hand, there is the understandable desire to let the student participate in the excitement offered by the many interesting new developments in all fields of physics. The discussion of such new topics gives the student an opportunity to see the actual growth process of science: new experiments, new techniques, and the attempts to relate new results to existing or emerging theoretical views. The study of the well established, traditional subjects of physics appears to lack these exciting aspects, and to offer little room for the display of creativity, except as historical fact.

It has at last been recognized that this need not entirely be so; that, in fact, the close ties between the traditional and the modern can be exploited to establish contacts between the classical subjects and current endeavors: classical mechanics and space navigation, wave optics and radar interferomery, or holography, astrophysical applications of classical electromagnetism and hydrodynamics, statistical mechanics as applied to biopolymers, or phase transitions, and so forth. To develop such links wherever they exist, and to put the essential parts of the traditional subjects into a modern perspective is an urgent and rewarding task.

On the undergraduate level, the recent burgeoning of such introductory texts as the Feynman lectures, the Berkeley physics course, and the Massachusetts Institute of Technology introductory physics texts bear witness to the interest that has been aroused by the problem of bringing the fundamentals of physics to the undergraduate in a novel way. This new series of MODERN PHYSICS MONOGRAPHS intends to continue this process at the more advanced level. It will present material for the upper level undergraduate and introductory graduate courses. At this level, there will be, on the one hand, courses of a specialized nature, with the purpose of giving the student an introduction to the great diversity of topics of physical science, from particle physics to nuclear, atomic, solid state, plasma and astrophysics, while, on the other hand, the traditional topics of the undergraduate sequences will be deepened and extended, and their interrelations more clearly established. We hope that the MODERN PHYSICS MONOGRAPHS series will help to give the lecturer in the field additional flexibility in choosing his course material and, if he is inclined to experiment, allow him to introduce into his course topics not generally covered in standard textbooks. In addition, the student will have access to a variety of collateral reading material.

For these very reasons, the books of this series are _not_ intended to be textbooks, but rather monographs; that is, works that cover a more restricted area in a space of approximately 200 to 400 pages. They contain problems with and without answers, and could either supplement existing texts or be used in groups as a replacement for a single text.

The present volume, a compact and lucid exposition of the basis structure of both classical and quantum mechanics, in their nonrelativistic version, is a case in point. The intriguing and profound parallels between classical and quantum-mechanical concepts (often given short shrift in "regular" texts), are here brought to the surface. The Feynman path integral, the Schwinger action principle, and their relation to Schroedinger's formulation are illuminated. Any student of classical and quantum mechanics will find this volume a storehouse of new insights and information.

Felix M. H. Villars

Cambridge, Massachusetts
January 1973

PREFACE

This book is based on a course that was designed to integrate the traditional graduate presentation of classical mechanics with an introductory treatment of quantum mechanics. For a number of years this course has been successfully given, in one quarter, to first year graduate students at UCLA. As prerequisites the student was expected to have taken undergraduate courses in both classical and quantum mechanics. Therefore, although the present book is introductory in nature, it is not intended as a first introduction to either classical or quantum mechanics.

The plan of the book is as follows. The first three chapters are devoted to general theory and the last two are concerned with applications. The first two chapters attempt to bring together the assumed prerequisite courses in classical and quantum mechanics in the framework of the underlying invariance group and its associated algebra. The development of the general theory that is begun in the first two chapters is worked out and concluded in the third. Chapter 4 deals with the motion of rigid bodies and Chapter 5 with the orbits of point particles. These last two chapters are tied together in terms of the general theory by making some use of the representation theory of the rotation group. The book is meant to be largely self-contained; however, the reader, depending on his preparation, may have additional work to do.

In combining the classical and quantum formalisms, greater weight must necessarily be given to quantum theory since that is the fundamental discipline. Classical theory unfortunately suffers in such an integration. In addition, although the relativistic generalization of classical mechanics would not be difficult and would be desirable at this point in the curriculum, the corresponding extension of quantum theory would present much more difficult questions, and we have accordingly decided to limit this book to nonrelativistic theory. Therefore there is less classical theory than one would perhaps like, but the clear advantages of an integrated course seem to us to be decisive. Finally a course in classical mechanics dealing with nonlinear mechanics and other modern topics would complete this program in a logical way.

I should like to thank my colleagues, Ernest Abers, Herbert Fried, David Saxon, and Roscoe White, as well as many students, particularly Joel Kvitky, John Mouton, Darwin Swanson, and Jan Smit, for helpful comments on the manuscript.

I should also like to thank Mr. Ronald Bohn for his skillful preparation of the typescript.

Robert J. Finkelstein

February, 1973
Los Angeles, California

Nonrelativistic Mechanics

CHAPTER 1

HAMILTONIAN FORMULATIONS (CLASSICAL THEORY)

1.1 INTRODUCTION

Most phenomena encountered in our normal experience appear to obey the laws of classical physics, but only because quantum effects are relatively unimportant in these situations. The exact description is always quantal, as far as we now know; and classical physics is only an approximate formalism which provides a good description in limiting situations according to the Bohr correspondence principle.

For example, it is believed that the motion of the earth about the sun is governed by quantum mechanics just as completely as the motion of an electron in the hydrogen atom.

In the planetary case, however, quantum corrections to the classical formulas are negligible while in the atomic case they are dominant. On the other hand one should not give the impression that quantum mechanics is important only in atomic and subatomic systems. There are well-known examples of macroscopic quantum systems: for example, superfluids, which are very cold, and white dwarf and neutron stars which are very dense.[1] As rough tests one may estimate h/S, the ratio of Planck's constant to an action which characterizes the phenomenon under consideration, or λ/R where λ is a quantum mechanical wave length and R is a characteristic linear dimension.

Quantum effects are important when these ratios are not small. For example

$$\frac{\lambda}{R} = \frac{h}{MVR}$$

is negligible when M and V refer to the mass and velocity of the earth, and R to its distance from the sun; but the same ratio becomes large in atoms. It also becomes large in macroscopic systems at low temperatures (V is small) at high densities (R is small) or even at normal densities and temperatures when M is small (electron gas in metals).

In this book an arbitrary physical system is described

in a realistic way, i.e., it is described classically when classical mechanics is adequate, and it is described by quantum theory when quantum effects are significant. More generally, the appropriate language always depends on the particular properties that we are trying to describe. Consider, for example, a mass of hydrogen or helium. If this system is very cold and completely frozen, then for some purposes it may be described as a rigid body with only six degrees of freedom. At a higher temperature it will be fluid and may be described hydrodynamically, and while liquid hydrogen is classical, liquid helium below the λ-point is superfluid and its behavior is dominated by quantum theory. If the flow becomes turbulent, one must take into account degrees of freedom that are neglected in the hydrodynamic description. If very large amounts of energy are introduced into the system, atomic and nuclear degrees of freedom become excited; then photons, and finally particles with rest mass, will be created. Clearly the appropriate language for describing one and the same system depends on circumstances.

By introducing more and more energy into any given system one may in principle generate all the different phenomena encountered in physics. We want to limit our considerations by confining our attention to non-relativistic systems: that means the exclusion of production and annihilation processes; it also means the exclusion of fields, since we shall be concerned with gravitational and electromagnetic forces only. In other words the subject of this book is simply a system of many particles which interact by instantaneous forces. This theoretical model is of enough generality to account for the main features of the physical world from the solar system down to atomic structure, and, after introducing forces of finite range, even down to much of nuclear structure. In the next section and the rest of the first chapter we shall discuss the classical description of such a system.

1.2 NON-RELATIVISTIC, CLASSICAL DESCRIPTION OF N-PARTICLE SYSTEM

Consider a many particle system, say a molecular gas, under conditions that permit us to neglect quantum and relativistic corrections.

The force on one particle then determines its motion according to the equation

$$M_i \frac{d^2 \underset{\sim}{x}_i}{dt^2} = \underset{\sim}{F}_i \tag{2.1}$$

where $\underset{\sim}{F}_i$ is the force on the ith particle due to all the other molecules in the gas. Equation (2.1) implies an inertial frame and Cartesian coordinates as well as a Galilean definition of mass and time.

We shall also make the simplest assumptions about the forces, namely,

(a) $\underset{\sim}{F}_i$ is the resultant of two body forces

$$\underset{\sim}{F}_i = \sum_{j \neq i} \underset{\sim}{F}_{ij} \tag{2.2}$$

(b) $\underset{\sim}{F}_{ij}$ is derivable from a two-body potential

$$\underset{\sim}{F}_{ij} = -\underset{\sim}{F}_{ji} = -\underset{\sim}{\nabla}_i V_{ij} \tag{2.3}$$

(c) The two-particle potential depends only on inter-particle distance

$$V_{ij} = V(|x_i - x_j|) \tag{2.4}$$

These forces include the Newtonian and Coulomb cases, which, except for small corrections, suffice for planetary and atomic physics, respectively. We exclude velocity and spin-dependence, etc.

Then, combining (2.1) - (2.4) one has

$$M_i \ddot{\underset{\sim}{x}}_i = -\underset{\sim}{\nabla}_i \left(\sum_{i \neq j} V_{ij} \; (|x_i - x_j|) \right) \tag{A)"}$$

Equations (A)" are a set of 3N equations for 3N functions

$x_i(t)$. To solve them one needs to assign positions and velocities at some given time, or an equivalent set of boundary conditions. We shall now consider other ways of writing (A)".

(a) Lagrangian Equations

Define the kinetic energy:

$$T = \frac{1}{2} \sum_i m_i \dot{\underset{\sim}{x}}_i^2 \tag{2.5}$$

the potential energy:

$$V = \sum_{i<j} V_{ij}(|\underset{\sim}{x}_i - \underset{\sim}{x}_j|) \tag{2.6}$$

the kinetic-potential, or the Lagrangian:

$$L = T - V \quad . \tag{2.7}$$

We shall also use the notation $x_{i\alpha}$ to indicate the α-coordinate of the ith particle when the vector notation is not convenient. Then by (2.5) - (2.7)

$$\frac{\partial L}{\partial x_{i\alpha}} = - \frac{\partial V}{\partial x_{i\alpha}} \qquad i = 1 \ldots N$$

$$\alpha = 1, 2, 3$$

$$\frac{d}{dt}\left(\frac{\partial L}{\partial \dot{x}_{i\alpha}}\right) = \frac{d}{dt}\left(\frac{\partial T}{\partial \dot{x}_{i\alpha}}\right) = m_i \ddot{x}_{i\alpha} \quad .$$

Therefore (A)" may also be written in the following form:

$$\frac{d}{dt}\left(\frac{\partial L}{\partial \dot{x}_{i\alpha}}\right) - \left(\frac{\partial L}{\partial x_{i\alpha}}\right) = 0 \qquad \begin{array}{l} i = 1 \ldots N \\ \alpha = 1, 2, 3 \end{array} \quad . \qquad \text{(A)}'$$

These are the Lagrangian equations.

The dynamical system is now characterized by the single function $L(x_1 \ldots x_N, \dot{x}_1 \ldots \dot{x}_N)$. It is also possible to make a point transformation from the Cartesian coordinates to any other set of 3N coordinates (q_k)

$$q_k = f_k(x_{1\alpha} \ldots x_{N\alpha} t) \qquad k = 1 \ldots 3N \quad .$$

As we shall see later, the equations (A)' then keep their same form

$$\frac{d}{dt}\left(\frac{\partial L}{\partial \dot{q}_k}\right) - \frac{\partial L}{\partial q_k} = 0 \qquad k = 1 \ldots 3N \quad . \qquad \text{(A)}$$

The (q_k) are usually referred to as generalized coordinates. The Lagrangian equations free us from dependence on Cartesian coordinates and permit us to solve problems for which other coordinates are more useful.[2]

Here we are interested in the general properties of the dynamical equations and not in the form they take in special coordinate systems. Nevertheless, we may continue to assume Cartesian coordinates without limiting the general discussion; but we shall sometimes denote the Cartesian coordinates by q_k.

(b) Hamiltonian Equations

The dynamical system under consideration is said to have 3N (= f) degrees of freedom. It may also be described by 2f equations of the first order instead of the f Lagrangian equations of the second order, as we shall now see. Define

$$p_k = \frac{\partial L}{\partial \dot{q}_k} \qquad k = 1\ldots f \quad , \tag{2.8}$$

$$H = \sum_k \dot{q}_k p_k - L \quad , \tag{2.9}$$

where the $\dot{q}_k$ appearing in H are supposed to be written as a function of the p_k and the q_k by (2.8). That is, in Cartesian coordinates,

$$p_{i\alpha} = m_i \dot{x}_{i\alpha} \quad . \tag{2.10}$$

One then uses (2.10) to eliminate $\dot{x}_{i\alpha}$ from (2.9):

$$\begin{aligned} H &= \sum p_{i\alpha}^2/m_i - \left(\sum p_{i\alpha}^2/2m_i - V \right) \\ &= \sum p_{i\alpha}^2/2m_i + V(x_{i\alpha}) \quad . \end{aligned} \tag{2.11}$$

Then

$$\frac{dx_{i\alpha}}{dt} = \frac{p_{i\alpha}}{m_i} = \frac{\partial H}{\partial p_{i\alpha}}$$

$$\frac{dp_{i\alpha}}{dt} = \frac{d}{dt}\frac{\partial L}{\partial \dot{x}_{i\alpha}} = \frac{\partial L}{\partial x_{i\alpha}} = -\frac{\partial V}{\partial x_{i\alpha}} = -\frac{\partial H}{\partial x_{i\alpha}} \quad .$$

The independent variables are now $x_{i\alpha}$ and $p_{i\alpha}$ and the corresponding equations of motion are

$$\frac{dx_{i\alpha}}{dt} = \frac{\partial H}{\partial p_{i\alpha}}$$

$$\frac{dp_{i\alpha}}{dt} = -\frac{\partial H}{\partial x_{i\alpha}} \quad . \qquad \text{(B)}'$$

Later it will be shown that these equations may also be written in terms of generalized coordinates

$$\frac{dq_k}{dt} = \frac{\partial H}{\partial p_k} \qquad k = 1\ldots f$$

$$\frac{dp_k}{dt} = -\frac{\partial H}{\partial q_k} \qquad f = 3N \quad . \qquad \text{(B)}$$

These are Hamilton's equations and the p_k and q_k are said to be canonically conjugate.

The preceding set (B) of 2f first order equations is equivalent to the set of f second order equations (A). In Chapter 2 it will be shown that these equations preserve their form under certain transformations of the following type:

$$Q_k = f_k\ (q_1\ldots p_1\ldots t) \qquad k = 1\ldots f$$

$$P_k = g_k\ (q_1\ldots p_1\ldots t) \quad . \qquad (2.12)$$

These equations include point transformations as special cases, but in general destroy the distinction between coordinates and momenta. For example, the following transformation, which turns out to be very important in quantum theory, clearly preserves the form of Hamilton's equations:

$$Q_k = -p_k$$

$$P_k = q_k \qquad . \tag{2.13}$$

The equations (2.12) are said to describe _contact_ transformations when they preserve the form of Hamilton's equations.

The product of two contact transformations is again a contact transformation. Such transformations combine according to the associative law, there is an identity, and every contact transformation has an inverse. A set of transformations with these properties is called a mathematical group.

In order that (2.12) describe a contact transformation, the functions f_k and g_k cannot be arbitrary but must satisfy certain conditions that are conveniently expressed in terms of so-called Poisson brackets. The Poisson bracket also provides a very direct connection between the classical and quantal formalism. We therefore introduce these objects next.

1.3 POISSON BRACKETS

Let $A(q_1 \ldots p_1 \ldots)$ and $B(q_1 \ldots p_1 \ldots)$ be any two dynamical functions. The Poisson bracket of A and B is defined as follows:

$$[A, B] = \sum_{k=1}^{f} \left[\frac{\partial A}{\partial q_k} \frac{\partial B}{\partial p_k} - \frac{\partial B}{\partial q_k} \frac{\partial A}{\partial p_k} \right] \tag{3.1}$$

where the sum is over all degrees of freedom.

The following properties are easily verified:

$$[A, B] = -[B, A] \tag{3.2}$$

$$[A, A] = 0 \tag{3.3}$$

$$[A, B + C] = [A, B] + [A, C] \tag{3.4}$$

$$[A, BC] = [A, B]C + B[A, C] \tag{3.5}$$

$$[A, [B, C]] + [B, [C, A]] + [C, [A, B]] = 0 \quad . \tag{3.6}$$

Equation (3.6) is known as the Jacobi identity.

The most elementary brackets are

$$[q_s, q_t] = [p_s, p_t] = 0 \tag{3.7}$$

$$[q_s, p_t] = \delta_{st} \tag{3.8}$$

$$[A, q_s] = \sum_{k} \left[\frac{\partial A}{\partial q_k} \frac{\partial q_s}{\partial p_k} - \frac{\partial q_s}{\partial q_k} \frac{\partial A}{\partial p_k} \right] \quad .$$

Therefore

$$[A, q_s] = -\frac{\partial A}{\partial p_s} \tag{3.9}$$

$$[A, p_s] = \frac{\partial A}{\partial q_s} \quad . \tag{3.10}$$

The time derivative of an arbitrary function $A(q_1 \ldots p_1 \ldots t)$ is

$$\frac{dA}{dt} = \sum_k \left(\frac{\partial A}{\partial q_k} \dot{q}_k + \frac{\partial A}{\partial p_k} \dot{p}_k \right) + \frac{\partial A}{\partial t} \quad .$$

By Hamilton's equations

$$\frac{dA}{dt} = \sum_k \left(\frac{\partial A}{\partial q_k} \frac{\partial H}{\partial p_k} - \frac{\partial A}{\partial p_k} \frac{\partial H}{\partial q_k} \right) + \frac{\partial A}{\partial t}$$

$$\frac{dA}{dt} = [A, H] + \frac{\partial A}{\partial t} \quad . \tag{3.11}$$

If A does not depend on the time explicitly,

$$\frac{dA}{dt} = [A, H] \quad . \tag{3.12}$$

If one puts $A = q_k$ or p_k in (3.12), then one recovers Hamilton's equations in P. B. form.

$$\dot{q}_k = [q_k, H] \tag{3.13a}$$

$$\dot{p}_k = [p_k, H] \tag{3.13b}$$

as one may check by comparing with (3.9) and (3.10).

If A does not contain t explicitly, and if

$$[A, H] = 0 \tag{3.14a}$$

then according to (3.11)

$$\frac{dA}{dt} = 0 \quad . \tag{3.14b}$$

In particular,

$$\frac{dH}{dt} = [H, H] = 0 \quad . \tag{3.15}$$

It was mentioned in (1.2) that Hamilton's equations (B) keep their form under contact transformations (2.12). The conditions for a contact transformation may be stated simply in terms of the P. B. as follows:

$$[Q_k, Q_\ell] = [P_k, P_\ell] = 0 \tag{3.16a}$$

$$[Q_k, P_\ell] = \delta_{k\ell} \tag{3.16b}$$

where

$$[Q_k, P_\ell] = \sum_{s=1}^{f} \left[\frac{\partial Q_k}{\partial q_s} \frac{\partial P_\ell}{\partial p_s} - \frac{\partial P_\ell}{\partial q_s} \frac{\partial Q_k}{\partial p_s} \right] \quad . \tag{3.16c}$$

These conditions will be established in Chapter 3. They may be regarded as partial differential equations for the functions Q and P as functions of the q and p.

With the aid of the P. B. one can express the equations of motion in the simple and general form (3.11) or (3.12). Of course one cannot, in general, solve the equations in this form any more than in the original Newtonian form (A)". It is now easier, however, to discuss the integrals of the motion and the symmetries on which these integrals depend. In fact, the only general simplification of the equations of motion depends on the existence of a symmetry or invariance group, and the P. B. is very well suited to a discussion of this group.

The elements of this group are contained in the wider group of all contact transformations. In general a contact transformation corresponds simply to a change in the mathematical description of a physical system, whereas the existence of a symmetry group depends on additional physical information or assumptions. This new input may be special to a particular system, or it may be common to all systems. For example, a particular system may be spherically symmetric; on the other hand, if the new input asserts that space itself is isotropic, then there is a simplification in the dynamical laws that govern all systems. In the latter case, even if the system is not spherically symmetric, the results of any experiment performed on it must be independent of the orientation of this system in space. In speaking of the invariance group we are, in the next section and in general, referring to symmetry of the dynamical laws that govern all systems.

1.4 THE INVARIANCE GROUP

If the physical system, A, is influenced by another system, B, then the physics of A depends on how far away it is from B, which then provides the origin of a privileged coordinate system; but if A is situated in otherwise empty space, one assumes that the dynamics of A is independent of where it is located. In other words, one assumes that space itself is homogeneous. Similarly, if A is influenced by an external force directed along a particular line, then the behavior of A will depend on its orientation with respect to that line; but if there is no physically distinguished direction in space, then one assumes that the dynamics of A is independent of the orientation of A in space. One says that space itself is isotropic. If both conditions hold, i.e., if there is neither a privileged point nor a privileged direction in space, then we assume that no observable property is changed by any rigid motion (translation or rotation) of the experimental laboratory.

Similarly if there is no privileged point in time either, one expects the laws of physics to be the same not only at all places but also at all times. Finally, according to the special principle of relativity, the dynamical equations are the same for all systems in uniform motion.[3] The various cases just described, namely rotations, translations, and boosts, or Galilean transformations, correspond to different possible changes in the laboratory and may be summarized as follows:

Rotations and Translations:

$$x_\alpha' = \sum_1^3 R_{\alpha\beta}\, x_\beta + a_\alpha \qquad \alpha = 1,2,3 \tag{4.1a}$$

$$p_\alpha' = \sum_1^3 R_{\alpha\beta}\, p_\beta \tag{4.1b}$$

Galilean Transformations:

$$x_\alpha' = x_\alpha - v_\alpha t \qquad \alpha = 1,2,3 \tag{4.2a}$$

$$t' = t + a_4 \tag{4.2b}$$

$$p_\alpha' = p_\alpha - m v_\alpha \qquad \alpha = 1,2,3 \tag{4.2c}$$

Equations (4.1) describe the changes in coordinates and momenta induced by a rigid displacement of the system or an opposite displacement of the coordinate frame. Equations

(4.2) give the transformation associated with a change from one inertial frame to another. The groups of all coordinate transformations associated with (4.1) and (4.2) are also known as the Euclidean and Galilean groups, respectively; the momentum transformations are not independent, but are determined by the coordinate transformations. The complete group has ten independent parameters.

In order to see how these transformations work out in more detail, consider first an arbitrary variation in the coordinates of all the particles: $\delta x_{i\alpha}$. The corresponding change in an arbitrary function of the coordinates and the conjugate momenta is then

$$\delta A = \sum_{i=1}^{N} \sum_{\alpha=1,2,3} \frac{\partial A}{\partial x_{i\alpha}} \delta x_{i\alpha} \quad . \tag{4.3}$$

We next specialize these coordinate changes in the following ways:

(a) Translations

If the $\delta x_{i\alpha}$ depend on the index i, then this mathematical variation describes a strain or deformation of the system. On the other hand, if the $\delta x_{i\alpha}$ are independent of i, that is, if

$$\delta x_{i\alpha} = \delta x_{\alpha} \tag{4.4}$$

then the system is rigidly translated by the amount $\delta\underset{\sim}{x}$. Then

$$\delta A = \sum_{\alpha} \delta x_{\alpha} \sum_{i=1}^{N} \frac{\partial A}{\partial x_{i\alpha}} \quad .$$

By (3.10)

$$\delta A = \sum_{\alpha} \delta x_{\alpha} \sum_{i=1}^{N} [A, p_{i\alpha}] \quad .$$

By (3.4)

$$\delta A = \sum_{\alpha} \delta x_{\alpha} [A, \sum_{i=1}^{N} p_{i\alpha}]$$

$$= \sum_{\alpha} \delta x_{\alpha} [A, P_{\alpha}] \tag{4.5a}$$

where

$$P_{\alpha} = \sum_{i=1}^{N} p_{i\alpha} = \sum_{i=1}^{N} m_i \dot{x}_{i\alpha} \tag{4.6}$$

or in vector notation

$$\delta A = \delta\underset{\sim}{x} [A, \underset{\sim}{P}] \quad . \tag{4.5b}$$

But if space is homogeneous, as we assume, the dynamics of an arbitrary system must be independent of such a translation. The Hamiltonian discussed in paragraph 2 is satisfactory in this respect. For

$$H = \sum \frac{\underset{\sim}{p}_i^{\,2}}{2m_i} + \sum_{i<j} V(|\underset{\sim}{x}_i - \underset{\sim}{x}_j|) \qquad (4.7)$$

and therefore H is unchanged by a translation:

$$\delta H = 0 \qquad (4.8a)$$

or by Eq. (4.5b)

$$[H, \underset{\sim}{P}] = 0 \qquad (4.8)$$

and by (3.14b)

$$\frac{d\underset{\sim}{P}}{dt} = 0 \quad . \qquad (4.9)$$

One may then say that translational invariance leads to the three integrals of linear momentum.

(b) Rotations

Consider the following special kind of displacements:

$$\begin{aligned} \delta x_{i1} &= -x_{i2}\,\delta\omega_i \\ \delta x_{i2} &= x_{i1}\,\delta\omega_i \quad . \end{aligned} \qquad (4.10)$$

Then the ith particle is rotated by $\delta\omega_i$ about the x_3-axis.

If $\delta\omega_i$ depends on the particle, the system will be subject to a twisting strain; but if $\delta\omega_i$ is the same for all particles, then again the physical system is rotated as a rigid body about the x_3-axis. On the other hand, if we wish to preserve the dynamical state, as well as the configuration, the momenta of all particles must be rotated by just the same amount. Hence a rigid rotation which preserves the dynamical state is characterized by

$$\begin{aligned} \delta x_{i1} &= -x_{i2}\,\delta\omega & \delta p_{i1} &= -p_{i2}\,\delta\omega \\ \delta x_{i2} &= x_{i1}\,\delta\omega & \delta p_{i2} &= p_{i1}\,\delta\omega \end{aligned} \tag{4.11}$$

and therefore instead of (4.3) one has

$$\begin{aligned} \delta A &= \sum_{i=1}^{N} \left[-\frac{\partial A}{\partial x_{i1}} x_{i2} + \frac{\partial A}{\partial x_{i2}} x_{i1} \right. \\ &\qquad\qquad \left. - \frac{\partial A}{\partial p_{i1}} p_{i2} + \frac{\partial A}{\partial p_{i2}} p_{i1} \right] \delta\omega \\ &= \delta\omega \sum_{i=1}^{N} \left[\left(x_{i1} \frac{\partial A}{\partial x_{i2}} - x_{i2} \frac{\partial A}{\partial x_{i1}} \right) \right. \\ &\qquad\qquad \left. + \left(p_{i1} \frac{\partial A}{\partial p_{i2}} - p_{i2} \frac{\partial A}{\partial p_{i1}} \right) \right] \\ &= \delta\omega \sum_{i=1}^{N} \{ x_{i1}[A,p_{i2}] + p_{i2}[A,x_{i1}] \\ &\qquad\qquad - x_{i2}[A,p_{i1}] - p_{i1}[A,x_{i2}] \} \end{aligned}$$

$$= \delta\omega \sum_{i=1}^{N} [A, x_{i1} p_{i2} - x_{i2} p_{i1}]$$

$$= \delta\omega [A, \sum_{i=1}^{N} (x_{i1} p_{i2} - x_{i2} p_{i1})]$$

or

$$\delta A = \delta\omega [A, L_{12}]$$

where

$$L_{12} = \sum_{i=1}^{N} (x_{i1} p_{i2} - x_{i2} p_{i1}) \quad .$$

In vector notation

$$\delta A = \delta\underset{\sim}{\omega} [A, \underset{\sim}{L}] \quad . \tag{4.12a}$$

But if space is isotropic, the Hamiltonian does not change when the system is rotated; this is indeed the case for the Hamiltonian (4.7). Therefore

$$\delta H = \delta\underset{\sim}{\omega} [H, \underset{\sim}{L}] = 0$$

or

$$[H, \underset{\sim}{L}] = 0 \tag{4.13}$$

and by (3.14)

$$\frac{d\underset{\sim}{L}}{dt} = 0 \quad . \tag{4.14}$$

One may say that rotational invariance leads to the three integrals of angular momentum.

(c) Galilean Transformations

According to Galilean relativity all inertial frames are equivalent. Although the Hamiltonian is no longer invariant,[4] the equations of motion are still unchanged when we transform from one inertial frame to another according to the following equations [see (4.2)]

$$\begin{aligned} \delta x_{i\alpha} &= -t\delta V_\alpha \\ \delta p_{i\alpha} &= -m_i \delta V_\alpha \quad . \end{aligned} \tag{4.15}$$

Then

$$\begin{aligned} \delta A &= \sum_{i\alpha} \left(\frac{\partial A}{\partial p_{i\alpha}} \delta p_{i\alpha} + \frac{\partial A}{\partial x_{i\alpha}} \delta x_{i\alpha} \right) \\ &= -\sum_\alpha \delta V_\alpha \sum_i \left(m_i \frac{\partial A}{\partial p_{i\alpha}} + t \frac{\partial A}{\partial x_{i\alpha}} \right) \\ &= \sum_\alpha \delta V_\alpha \; [A, \sum_i (m_i x_{i\alpha} - t p_{i\alpha})] \end{aligned}$$

or

$$\delta A = \delta \underset{\sim}{V} \cdot [A, \underset{\sim}{G}] \tag{4.16}$$

where

$$\underset{\sim}{G} = \sum_i (m_i \underset{\sim}{x}_i - t \underset{\sim}{p}_i) \quad . \tag{4.17}$$

Furthermore, if the system is free,

$$\frac{d\underset{\sim}{G}}{dt} = \sum_i m_i \dot{\underset{\sim}{x}}_i - \underset{\sim}{P} = 0 \quad . \tag{4.18}$$

Notice that

$$\frac{d\underset{\sim}{G}}{dt} = \frac{\partial \underset{\sim}{G}}{\partial t} + [\underset{\sim}{G}, H] = 0 \tag{4.19}$$

but

$$[\underset{\sim}{G}, H] \neq 0 \qquad \text{unless} \qquad \underset{\sim}{P} = 0 \quad .$$

Another way to express $\underset{\sim}{G}$ is

$$\underset{\sim}{G} = M\underset{\sim}{X} - t\underset{\sim}{P} \tag{4.20}$$

where $\underset{\sim}{X}$ is the center of mass.[5]

Just as the rotation-translation group implies that there are no distinguished points or lines in physical space, the Galilean group implies that there are no privileged states of motion or no distinguished inertial frames. The corresponding conservation law is

$$\frac{d\underset{\sim}{G}}{dt} = 0 \tag{4.21}$$

or

$$\underset{\sim}{X} = \frac{t}{M}\underset{\sim}{P} + \text{const.}, \tag{4.22}$$

which states that the center of mass of an isolated system moves uniformly in a straight line (in an inertial frame).

1.5 ALGEBRA OF GENERATORS

In the preceding section the following equations were obtained:

$$\frac{dA}{dt} = [A, H] + \frac{\partial A}{\partial t} \tag{5.1}$$

$$\frac{dA}{dx} = [A, P_x] \tag{5.2}$$

$$\frac{dA}{d\omega_{xy}} = [A, L_{xy}] \tag{5.3}$$

$$\frac{dA}{dV_x} = [A, G_x] \quad . \tag{5.4}$$

In view of these results one may describe H and $\underset{\sim}{P}$ as the generators of time and space displacements respectively; similarly $\underset{\sim}{L}$ and $\underset{\sim}{G}$ are the generators of rotations and Galilean transformations respectively. Altogether there are ten generators corresponding to the ten parameters of the complete group.

If A is the Hamiltonian of an isolated system, then it is invariant under rotations, translations, and time displacements. Therefore

$$[H, P_x] = [H, L_{xy}] = \frac{\partial H}{\partial t} = 0 \quad . \tag{5.5}$$

But if A is an arbitrary function of the dynamical variables, the corresponding derivatives and brackets do not vanish. For example, let X be the center of mass vector:

$$\underset{\sim}{X} = \frac{\sum m_i \underset{\sim}{x}_i}{\sum m_i} \quad . \tag{5.6}$$

Then

$$[X_r, P_s] = \delta_{rs} \tag{5.7}$$

$$[X_r, H] = P_r/M \quad . \tag{5.8}$$

In general, if A is an arbitrary tensor, then $\delta A/\delta\omega_{xy}$ is entirely determined by its tensor transformation law. For example, the vector $\underset{\sim}{A}$ transforms as follows:

$$\delta A_x = -A_y \; \delta\omega_{xy}$$

$$\delta A_y = A_x \; \delta\omega_{xy} \quad .$$

If A_x is now put in (5.3), one has

$$-A_y \ \delta\omega_{xy} = [A_x, L_{xy}] \ \delta\omega_{xy}$$

or

$$[A_x, L_z] = -A_y$$

and

$$[L_s, A_t] = \varepsilon_{stm} A_m \quad , \tag{5.9}$$

where ε_{stm} is completely antisymmetric and $\varepsilon_{123} = 1$. (The repeated index is always summed here.) This relation includes the case $\underset{\sim}{A} = \underset{\sim}{L}$. Then

$$[L_s, L_t] = \varepsilon_{stm} L_m \quad . \tag{5.10}$$

The Poisson bracket of higher rank tensors may be obtained in just the same way.

The complete set of P. B. relations for all the generators of the non-relativistic symmetry group is given in the following table:

	P_ℓ	G_ℓ	L_ℓ	H
P_k	0	$-M\,\delta_{k\ell}$	$\varepsilon_{k\ell m}\,P_m$	0
G_k		0	$\varepsilon_{k\ell m}\,G_m$	P_k
L_k			$\varepsilon_{k\ell m}\,L_m$	0
H				0

(5.11)

The generators of the rotation-translation subgroup form a Lie algebra[6] and satisfy relations of the following kind:

$$[X_k,\ X_\ell] = C_{k\ell m}\ X_m \tag{5.12}$$

where the $C_{k\ell}{}^{m}$ are called structure constants.[7] They are independent of time. The generators are also independent of time.

$$\frac{dX_k}{dt} = 0 \quad . \tag{5.13}$$

Furthermore, the generators of the rotation-translation subgroup all satisfy the equations

$$[X_k,\ H] = 0 \quad . \tag{5.14}$$

Equations (5.12) and (5.14) are consistent only if

$$[[X_k,\ X_\ell],\ H] = 0 \tag{5.15}$$

but this relation is guaranteed by (5.14) and the Jacobi identity (3.6), which now reads

$$[H[X_k, X_\ell]] + [X_k[X_\ell, H]] + [X_\ell[H, X\ ]] = 0 \quad . \qquad (5.16)$$

NOTES ON CHAPTER 1

1. Evidence continues to grow that the present universe originated from a very condensed state ("big bang"). If this view is correct, then quantum theory in conjunction with general relativity probably plays a fundamental role in questions of cosmology. The same remark must be made about the conjectured "gravitational collapse" of a stellar mass.
2. The generalized coordinates no longer necessarily resolve the system into individual particles.
3. According to special relativity, the dynamical equations are the same only for systems in uniform motion with respect to each other. This class of privileged observers is called "inertial".
4. The Hamiltonian is not a scalar under Galilean transformations; and it is one component of the energy-momentum vector under a Lorentz transformation.

5. In the Lorentz group the three components of $\underset{\sim}{G}$ and the three components of $\underset{\sim}{L}$ are combined into the six generators of a four-dimensional rotation. That is

$$G_k = MX_k - tP_k$$

is the limit of

$$M_{ok} = EX_k - tP_k$$

and the conservation of M_{ok} reads as follows

$$\frac{dM_{ok}}{dt} = 0 \qquad \text{and therefore} \qquad \frac{dX_k}{dt} = \frac{P_k}{E}$$

when applied to the motion of an isolated system. The six generators of a four-dimensional rotation consist of M_{ok} and M_{ik} (= L_{ik}).

6. See, for example, reference 7, page 301.

   The generators X_k in (5.12) and the following equations of course have nothing to do with the same symbol used earlier for center of mass.

7. The bracket $[P_k, G_\ell]$ does not satisfy (5.12). The corresponding representation is called projective. See, for example, reference 7, page 458. However, the replacement of G_k by M_{ok} also changes the bracket:

$$[P_k, G_\ell] \to [P_k, M_{o\ell}] = [P_k, EX_\ell - tP_\ell]$$
$$= -E\delta_{k\ell} = -P_o\delta_{k\ell}$$

where P_o is the fourth component of the energy momentum vector. Then all Poisson brackets satisfy (5.12). The corresponding algebra belongs to the inhomogeneous Lorentz group, also called the Poincaré group.

BIBLIOGRAPHY FOR CHAPTER 1

General Works on Classical Dynamics

1. E. T. Whittaker, A Treatise on the Analytical Dynamics of Particles and Rigid Bodies, Dover (1944).
2. L. D. Landau and E. M. Lifshitz, Mechanics, Addison-Wesley (1960).
3. H. Goldstein, Classical Mechanics, Addison-Wesley (1950).
4. H. C. Corben and P. Stehle, Classical Mechanics, Wiley (1960).
5. G. J. Konopinski, Classical Descriptions of Motion, Freeman (1969).
6. J. Marion, Classical Dynamics of Particles and Systems, Academic Press (1965).

Group Theory

7. M. Hammermesh, Group Theory and Its Applications to

Physical Problems, Addison-Wesley (1962).

PROBLEMS

1. Show that Lagrange's equations are form invariant under point transformations.

2. Prove

   a) $[A, BC] = [A, B]C + B[A, C]$

   b) $[A,[B, C]] + [B,[C, A]] + [C[A, B]] = 0$.

3. Show that the following are contact transformations

   a) $Q_k = aq_k + bp_k$ $\quad \begin{vmatrix} a & b \\ c & d \end{vmatrix} = 1$

   $P_k = cq_k + dp_k$

   b) $Q = \ln\left(\frac{1}{q} \sin p\right)$

   $P = q \cot p$

   c) $Q_k = \tan^{-1} (q_k/p_k)$

   $P_k = \frac{1}{2} (p_k{}^2 + q_k{}^2)$

4. Find how H transforms under Galilean transformations. Show that Hamilton's equations are invariant under Galilean transformations.

5. Show that angular momentum in an arbitrary inertial frame is

$$\underset{\sim}{L} = \underset{\sim}{L}^o + \underset{\sim}{R} \times \underset{\sim}{P}$$

where $\underset{\sim}{L}^{o}$ is the angular momentum with respect to the center of mass, and $\underset{\sim}{R}$ is the radius vector of the center of mass, while $\underset{\sim}{P}$ is the total momentum of the system. Show how angular momentum transforms under translations and Galilean transformations. Check by using Table (5.11) and (4.5) and (4.16).

6. Show

$$[X_\alpha, P_\beta] = \delta_{\alpha\beta}$$

$$[X_\alpha, H] = \frac{1}{M} P_\alpha$$

where X_α is the center of mass vector.

7. Find

$$[A_{MN}, L_{KS}]$$

where A_{MN} is an antisymmetric tensor.

8. Find the rate of change of energy and momentum of a system with the following Hamiltonian:

$$H = \overset{o}{H} + V(\underset{\sim}{x}, t)$$

where $\overset{o}{H}$ is independent of $\underset{\sim}{x}$ and t.

9. The Hamiltonian of a harmonic oscillator is

$$H = \frac{1}{2m} (P^2 + m^2\omega^2 q^2) \quad .$$

Write Hamilton's equations. Find the Lagrangian and write Lagrange's equations.

10. Write the Lagrangian and Hamiltonian of a coplanar double pendulum in terms of the coordinates (ℓ_1, ℓ_2, ϕ_1, ϕ_2)

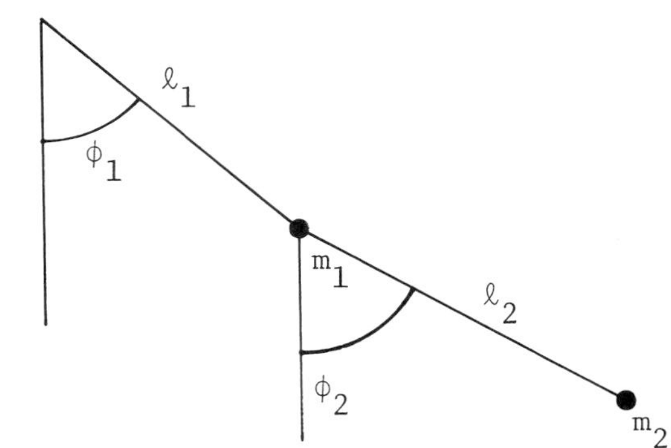

11. Consider the description of a single particle in a central potential, V(r). First introduce spherical polar coordinates r, θ, φ:

$$x = r \sin\theta \cos\phi$$

$$y = r \sin\theta \sin\phi$$

$$z = r \cos\theta \quad .$$

Write the Lagrangian in these coordinates. What are Lagrange's equations? Find the canonical momenta p_r, p_θ, p_ϕ . Write the Hamiltonian and the three Hamilton's equations. What is the physical significance of p_θ? Compute [p , H].

12. Consider a charged particle in an electromagnetic field

described by a scalar potential $\phi(\underset{\sim}{r})$ and a vector potential $\underset{\sim}{A}(\underset{\sim}{r})$. The electric and magnetic fields are

$$\underset{\sim}{E}(\underset{\sim}{r}) = -\underset{\sim}{\nabla}\phi(\underset{\sim}{r}) - \frac{1}{c}\frac{\partial \underset{\sim}{A}(\underset{\sim}{r})}{\partial t}$$

$$\underset{\sim}{B}(\underset{\sim}{r}) = \underset{\sim}{\nabla} \times \underset{\sim}{A}(\underset{\sim}{r}) \quad .$$

Show that the Lagrangian

$$L = \frac{1}{2} m \dot{\underset{\sim}{r}}^2 - q\,\phi(\underset{\sim}{r}) + \frac{q}{c}\underset{\sim}{A}(\underset{\sim}{r})\,\dot{\underset{\sim}{r}}$$

gives rise to the Lorentz force law. What is the canonical momentum $\underset{\sim}{p}$? What is the Hamiltonian $H(\underset{\sim}{p},\underset{\sim}{r})$? Calculate the Hamiltonian equations of motion.

13. Dilatation.

Define $\quad D = \sum \underset{\sim}{x}_i\,\underset{\sim}{p}_i$

Show that $[A,D] = nA$

if A is a homogeneous function of the coordinates, i.e., if

$$A(\lambda x_{i\alpha}) = \lambda^n A(x_{i\alpha}) \quad .$$

14. Find the Poisson brackets between $\underset{\sim}{P}$, H, $\underset{\sim}{L}$, $\underset{\sim}{G}$, and D. Does the algebra close if $\underset{\sim}{G}$ is replaced by M_{ok}?

15. Virial Theorem.

Show that the change in D during the time, T, is

$$D(T) - D(0) = \sum_{i\alpha} \int_0^T \left(p_{i\alpha} \frac{\partial H}{\partial p_{i\alpha}} - x_{i\alpha} \frac{\partial H}{\partial x_{i\alpha}} \right) dt \quad .$$

If $x_{i\alpha}$ and $p_{i\alpha}$ remain bounded, so does D, and therefore

$$\lim_{T\to\infty} \left[\frac{D(T) - D(0)}{T} \right] = 0$$

so that the time average of the right side also vanishes:

$$\sum_{i\alpha} < p_{i\alpha} \frac{\partial H}{\partial p_{i\alpha}} > = \sum_{i\alpha} < x_{i\alpha} \frac{\partial H}{\partial x_{i\alpha}} >$$

or

$$<2T> = \sum_{i\alpha} < x_{i\alpha} \frac{\partial V}{\partial x_{i\alpha}} >$$

$$= - < \sum_i \underset{\sim}{x}_i \underset{\sim}{F}_i > \quad .$$

This is the virial theorem. If the potential energy is itself homogeneous of degree k, then

$$<2T> = k <U>$$

and also

$$<T> = \frac{k}{k+2} E \quad .$$

16. A special case of (5.12) is

$$[G_m, L_n] = \varepsilon_{mnk} G_k \quad .$$

In contrast to (5.14), however, $[G_k, H] \neq 0$. Then

$$[[G_m, L_n], H] = \varepsilon_{mnk}[G_k, H] = \varepsilon_{mnk}P_k \quad .$$

Evaluating the left side by the Jacobi identity, show that this equation is satisfied.

CHAPTER 2

HAMILTONIAN FORMULATIONS (QUANTUM THEORY)

2.1 THE HEISENBERG EQUATIONS OF MATRIX MECHANICS

According to classical mechanics one may in principle make a complete measurement of all the coordinates and conjugate momenta at any given time, but according to quantum mechanics the simultaneous determination of the p's and q's is forbidden by the Heisenberg uncertainty principle. Classical mechanics then predicts from the given initial data the unique evolution of any physical system such as the N-particle gas that we have been discussing. On the other hand, according to quantum mechanics, one is able to calculate only the probability of a particular evolution from the given initial conditions. Although many of the quantal ways of evolving are close to the unique classical possibility, most are not; and in particular situations quantum theory predicts, for example, tunneling, superfluidity and other phenomena that are classically incomprehensible. All of this is of course not surprising if we already understand that classical mechanics is only a limiting case of quantum mechanics.

The earliest forms of quantum mechanics to be discovered were known as matrix mechanics (Heisenberg) and wave mechanics (Schrödinger) and at first it was not realized that these were equivalent. Other important formulations were subsequently discovered by Dirac, Feynman, and Schwinger. We shall discuss all of these formulations since they give different and complementary insights into quantum mechanics.

We shall first adopt the Dirac approach in order to go over from the classical description of the system (the nonrelativistic gas) already discussed in the first chapter to the corresponding quantum description. In order to accomplish this, define the commutator of two matrices (A, B) as follows:

$$(A, B) = AB - BA \quad . \tag{1.1}$$

The commutator has the same algebraic properties as the Poisson bracket, namely:

$$(A, B) = -(B, A) \tag{1.2}$$

$$(A, A) = 0 \tag{1.3}$$

$$(A, B + C) = (A, B) + (A, C) \tag{1.4}$$

$$(A, BC) = (A, B)C + B(A, C) \tag{1.5}$$

$$(A, (B, C)) + (B, (C, A)) + (C, (A, B)) = 0 \quad . \tag{1.6}$$

For example

$$\begin{aligned}(A, BC) &= A\ BC - BC\ A \\ &= (AB - BA)C + B(AC - CA) \\ &= (A, B)C + B(A, C) \quad .\end{aligned}$$

In view of these relations one can without algebraic inconsistency substitute in the classical equations as follows:

$$[\ ,\ ] \to a^{-1}(\ ,\ ) \tag{1.7}$$

where a is an arbitrary constant, to obtain a new set of relations involving commutators of matrices. By section (1.3)

$$\frac{dA}{dt} = \frac{(A, H)}{a} + \frac{\partial A}{\partial t} \tag{1.8}$$

$$\frac{\partial A}{\partial q_i} = \frac{(A, p_i)}{a} \tag{1.9}$$

$$\frac{\partial A}{\partial p_i} = -\frac{(A, q_i)}{a} \quad . \tag{1.10}$$

Then also

$$(q_s, p_r) = a\delta_{sr} \tag{1.11}$$

$$(q_s, q_r) = (p_s, p_r) = 0 \quad . \tag{1.12}$$

If the matrices representing the q_k and the p_k are hermitian, as one requires in quantum mechanics, then it follows from (1.11) that a is pure imaginary [see remark after (A.12) in Appendix A]. We therefore put

$$a = i\hbar \quad . \tag{1.13}$$

If $\hbar$ is now identified with Planck's constant, the equations just written are the matrix equations that were first discovered by Heisenberg. The foregoing way of relating the classical and quantal relations, with the aid of the correspondence between the Poisson brackets and the commutators, was discovered by Dirac.

We shall now study the equations of matrix mechanics in the following form:

$$\frac{dA}{dt} = \frac{(A, H)}{i\hbar} + \frac{\partial A}{\partial t} \tag{1.14}$$

$$\frac{\partial A}{\partial q_i} = \frac{(A, p_i)}{i\hbar} \tag{1.15}$$

$$-\frac{\partial A}{\partial p_i} = \frac{(A, q_i)}{i\hbar} \quad . \tag{1.16}$$

The canonical commutators (1.11) and (1.12) are contained in (1.15) and (1.16). The corresponding equations connecting the generators of the invariance group are obtained from (5.12) of Chapter 1 by the substitution (1.7)

$$(X_k, X_\ell) = i\hbar \, C_{k\ell}{}^m \, X_m \tag{1.17}$$

while equations like (5.2) - (5.4) become

$$\frac{\delta A}{\delta \xi_m} = \frac{(A, X_m)}{i\hbar} \tag{1.18}$$

where $\delta\xi_m$ is the kind of displacement induced by X_m.

The formal relation of the matrix equations to the corresponding classical relations may be further clarified by the following remark. Let A(p q) and B(p q) be two polynomials. Then it may be shown that

$$\lim_{\hbar \to 0} \frac{(A, B)}{i\hbar} = [A, B] \tag{1.19}$$

if the canonical commutation rules are satisfied. The right side of this equation is the Poisson bracket. Equation (1.19) is a statement of Bohr's correspondence principle when properly interpreted.

The Dirac method of going over from the classical to the

quantum equations must be regarded as a recipe, or a mnemonic, and certainly not as a derivation; since a derivation should go in the reverse direction, from quantum to classical theory in the general way indicated by (1.19). One might also object that we are here presenting the quantum equations before explaining the meaning of the symbols appearing in the formalism; and it is true that we are describing the mathematics before the physics. This physically illogical order was, however, actually followed during the historical development of the quantum theory, insofar as Heisenberg and Schrödinger proposed their respective formalisms before the currently accepted probability interpretation was discovered by Born.

2.2 PHYSICAL INTERPRETATION[1]

In the equations of matrix mechanics every physical observable is represented by a matrix. For example, the position is described by a matrix vector $||\underset{\sim}{X}_{nm}||$ instead of a numerical vector $\underset{\sim}{x}$, and the orbit is somehow related to $||\underset{\sim}{X}_{nm}(t)||$ instead of to $\underset{\sim}{x}(t)$. The matrix entries in $||\underset{\sim}{X}_{nm}(t)||$ are rather arbitrary[2] as will be discussed further in 2.5.

The eigenvalues and eigenvectors of these matrices are not arbitrary, however, and they are the basis of the physical interpretation that we shall now describe.

Physical Postulates

Every physical observable (A) is represented in the theory by a hermitian matrix (A) and the only possible results of a measurement of A are its eigenvalues.

Consider an arbitrary physical system. We assume that every possible state (S) of the system is represented by a ray (a vector undetermined up to a phase factor) in the vector space spanned by the orthonormal eigenvectors of A as follows:

$$\psi_S = \sum_{a'} c_S(a')\ \psi(a') \tag{2.1}$$

where ψ_S is the ray representing S and where $\psi(a')$ is the eigenvector belonging to the eigenvalue a'. Here ψ_S is called the state vector and is assumed to completely describe the state S of the quantum system. (If the spectrum of A contains an infinite number of eigenvalues, then the vector space spanned by the eigenvectors of A is ∞-dimensional.)

The probability of obtaining the particular result a' when an observation of A is made on the physical system in state ψ_S is

$$P_S(a') = |c_S(a')|^2 \tag{2.2}$$

where $c_S(a')$ is the component of ψ_S in the $\psi(a')$ direction.

Finally

$$\sum_{a'} |c_S(a')|^2 = \sum_{a'} P_S(a') = 1 \qquad (2.3)$$

since the a' are supposed to completely label all possible outcomes of a measurement of A.

2.3 INTERPRETATION FORMULATED IN DIRAC BRACKET NOTATION

The Dirac notation for quantum mechanics is so powerful that it has become almost an essential part of the subject. We shall now describe this notation.

(a) Expansion (Superposition Principle)[3]

Equation (2.1) may be expressed in the following abbreviated form:

$$|S\rangle = \sum_{a'} c_S(a')|a'\rangle \qquad (3.1)$$

where

$$\psi_S = |S\rangle \qquad (3.2)$$

$$\psi(a') = |a'\rangle \quad . \qquad (3.3)$$

The eigenvalues and eigenvectors are defined by the equation

$$A|a'\rangle = a'|a'\rangle \qquad (3.4)$$

which is an abbreviation of

$$\sum_{s} A_{rm}\,\psi_m(a') = a'\,\psi_r(a') \quad , \tag{3.4a}$$

and $\psi_m(a')$ is the mth component of the vector representing the physical state a'. The complex conjugate of this equation is

$$\sum_{m} \psi_m(a')^*\,A^*_{rm} = (a')^*\,\psi_r(a')^*$$

or

$$\sum_{m} \psi_m(a')^*\,A^{\dagger}_{mr} = (a')^*\,\psi_r(a')^* \tag{3.5a}$$

where $A^{\dagger}$ is the hermitian adjoint, that is, $A^{\dagger}_{mr} = A^*_{rm}$.

Equation (3.5a) may now be written in bracket notation as follows:

$$\langle a'|A^{\dagger} = \langle a'|(a')^* \tag{3.5}$$

where $\psi(a')^*$ standing on the left becomes $\langle a'|$. If A represents a physical observable and is therefore an hermitian matrix, (3.5) becomes

$$\langle a'|A = \langle a'|a' \quad . \tag{3.5b}$$

The inner product of $\langle a'|$ and $|b'\rangle$ is now denoted by

$$\langle a'|b'\rangle = \sum_m \psi_m(a')^* \psi_m(b') \quad . \tag{3.6}$$

$\langle a'|b'\rangle$ is the Dirac bracket. $\langle a'|$ and $|b'\rangle$ are called bra and ket vectors.

In the Appendix it is shown for hermitian and unitary operators that $\langle a''|a'\rangle = 0$ unless $a'' = a'$. We may also normalize, so that

$$\langle a''|a'\rangle = \delta(a'', a') \quad . \tag{3.7}$$

One may then solve (3.1) for the expansion coefficients

$$\begin{aligned} \langle a''|S\rangle &= \sum c_S(a') \langle a''|a'\rangle \\ &= c_S(a'') \quad . \end{aligned} \tag{3.8}$$

The expansion itself may then be written entirely in bracket notation:

$$|S\rangle = \sum_{a'} |a'\rangle \langle a'|S\rangle \quad . \tag{3.9}$$

For an unlabeled state it reads

$$|\ \rangle = \sum_{a'} |a'\rangle \langle a'|\ \rangle \quad . \tag{3.10}$$

If $|b'\rangle$ is an eigenvector of another observable (B), then

$$|b'\rangle = \sum |a'\rangle \langle a'|b'\rangle \quad . \tag{3.11}$$

This equation may be multiplied on the left by $\langle c'|$

$$\langle c'|b'\rangle = \sum_{a'} \langle c'|a'\rangle \langle a'|b'\rangle \tag{3.12}$$

to give the composition law of the brackets.

The following properties should also be noticed

$$\langle a'|b'\rangle^* = \langle b'|a'\rangle \tag{3.13}$$

$$\sum_{a'} |a'\rangle \langle a'| = 1 \quad . \tag{3.14}$$

(3.14) follows from (3.10) or (3.12), and is known as the completeness condition.

(b) Change of Basis

The transformation from one orthonormal basis to another is unitary.[4] That is, if

$$\langle a'|a''\rangle = \delta(a', a'')$$

$$\langle b'|b''\rangle = \delta(b', b'')$$

then

$$\langle a''|b'\rangle = \langle a''|U|a'\rangle$$

is a unitary matrix, i.e.,

$$U^\dagger = U^{-1}$$

where $\dagger$ again means hermitian adjoint. Then

$$|b''\rangle = \sum |a'\rangle \langle a'|U|a''\rangle$$
$$= U\;|a''\rangle$$

by (3.11) and (3.14). The adjoint of this last equation is $\langle b''| = \langle a''|\; U^{-1}$, and therefore

$$\langle b'|X|b''\rangle = \langle a'|U^{-1}\; X\; U|a''\rangle \qquad .$$

In particular

$$\langle b'|U|b''\rangle = \langle a'|U|a''\rangle \qquad .$$

Equation (3.12) states that the product of two unitary transformations is aso unitary. Geometrically the matrix $\langle a''|b'\rangle$ may be thought of in terms of generalized direction cosines between the old and the new basis.

(c) Probabilities

In terms of the bracket notation one may say that the probability of observing the eigenvalue a' for the system in state S is

$$P_S(a') = |\langle a'|S\rangle|^2 \qquad . \tag{3.15}$$

Therefore the transformation coefficient $\langle a''|b'\rangle$ is the probability amplitude of the result a" if the system is prepared to be in the state $|b'\rangle$. The normalization condition

$$\sum P_S(a') = 1$$

becomes

$$\begin{aligned}
\langle S|S\rangle &= \sum_{a'} \langle S|a'\rangle \langle a'|S\rangle \\
&= \sum_{a'} |\langle a'|S\rangle|^2 \\
&= \sum P_S(a') \\
&= 1
\end{aligned}$$

or simply

$$\langle S|S\rangle = 1 \quad . \tag{3.16}$$

(d) Expectation Values

The expectation value of A, for example, when the system is in the state S, is

$$\langle A\rangle = \langle S|A|S\rangle \quad . \tag{3.17}$$

Proof:

$$\langle S|A|S\rangle = \sum_{a'a''} \langle S|a'\rangle \langle a'|A|a''\rangle \langle a''|S\rangle \quad .$$

But A is diagonal in the $|a'\rangle$ basis. Therefore

$$\begin{aligned}
\langle S|A|S\rangle &= \sum_{a'} \langle S|a'\rangle\, a'\, \langle a'|S\rangle \\
&= \sum_{a'} a'\, |\langle a'|S\rangle|^2
\end{aligned}$$

$$= \sum_{a'} a' \, P_S(a')$$

which is the usual expectation value.

(e) Dispersion

If the system is an eigenstate $|a'\rangle$, then a measurement of A is certain to lead to the corresponding eigenvalue. Otherwise a repetition of the experiment will give a collection of values a', all eigenvalues of A. The dispersion of this collection of values is given by

$$\langle(\Delta A)^2\rangle = \langle(A - \langle A\rangle)^2\rangle = \langle A^2\rangle - 2\langle A\rangle\langle A\rangle + \langle A\rangle^2$$
$$= \langle A^2\rangle - \langle A\rangle^2 \quad . \tag{3.18}$$

Suppose that the system is in an eigenstate of B:

$$|S\rangle = |b'\rangle \quad .$$

Then

$$\langle b'|(\Delta A)^2|b'\rangle = \langle b'|A^2|b'\rangle - \langle b'|A|b'\rangle^2 \quad .$$

But

$$\langle b'|A^2|b'\rangle = \sum_{b''} \langle b'|A|b''\rangle \langle b''|A|b'\rangle$$
$$= |\langle b'|A|b'\rangle|^2 + \sum_{b''\neq b'} |\langle b'|A|b''\rangle|^2 \quad .$$

Therefore

$$\langle(\Delta A)^2\rangle = \langle b'|(\Delta A)^2|b'\rangle = \sum_{b''\neq b'} |\langle b'|A|b''\rangle|^2 \qquad (3.19)$$

for the state b'. Higher moments may be obtained by a similar procedure.

Therefore the off-diagonal elements measure the dispersion. If A and B commute, then they may be taken diagonal simultaneously and both measurements are then simultaneously free from dispersion.

According to the interpretation just described, the orbit of a single particle is approximately determined by $\langle S|X(t)|S\rangle$ and the statistical uncertainty in the orbit is determined by the off-diagonal elements of X(t) or higher moments such as $\langle S|X^2(t)|S\rangle$. In the classical limit the orbit may become very sharp with negligible dispersion. However, to discuss the positional coordinate in more detail we need a procedure for dealing with observables having a continuous spectrum. This procedure will be described in paragraph 2.5.

(f) Example

We now consider the most unclassical conditions, or the extreme quantum limit, since this situation provides the simplest illustration of the quantum formalism. For example consider an atom whose angular momentum is definitely $\frac{1}{2}\hbar$.

Then the angular momentum vector of the atom is

$$\underset{\sim}{S} = \frac{1}{2}\hbar\, \underset{\sim}{\sigma} \tag{3.20}$$

where

$$\sigma_x = \begin{pmatrix} 0 & 1 \\ 1 & 0 \end{pmatrix} \qquad \sigma_y = \begin{pmatrix} 0 & -i \\ i & 0 \end{pmatrix} \qquad \sigma_z = \begin{pmatrix} 1 & 0 \\ 0 & -1 \end{pmatrix} \tag{3.21}$$

The three components of $\underset{\sim}{S}$ of course satisfy the commutation rules of an angular momentum. These are contained in (1.17) and are

$$(S_x, S_y) = i\hbar S_z \quad .$$

Then

$$S_x^{\ 2} + S_y^{\ 2} + S_z^{\ 2} = \frac{3}{4}\hbar^2 = \frac{1}{2}\left(\frac{1}{2} + 1\right)\hbar^2 = S(S+1)\hbar^2$$

where $S = 1/2$. The eigenvectors of S_z are $\begin{pmatrix} 1 \\ 0 \end{pmatrix}$ and $\begin{pmatrix} 0 \\ 1 \end{pmatrix}$ since

$$\begin{aligned} S_z \begin{pmatrix} 1 \\ 0 \end{pmatrix} &= \frac{\hbar}{2}\begin{pmatrix} 1 \\ 0 \end{pmatrix} \\ S_z \begin{pmatrix} 0 \\ 1 \end{pmatrix} &= -\frac{\hbar}{2}\begin{pmatrix} 0 \\ 1 \end{pmatrix} \quad . \end{aligned} \tag{3.22}$$

Similarly, the normalized eigenvectors of S_x are $\frac{1}{\sqrt{2}}\begin{pmatrix} 1 \\ 1 \end{pmatrix}$ and $\frac{1}{\sqrt{2}}\begin{pmatrix} 1 \\ -1 \end{pmatrix}$ since

$$S_x \begin{pmatrix} 1 \\ 1 \end{pmatrix} = \frac{\hbar}{2} \begin{pmatrix} 1 \\ 1 \end{pmatrix}$$

$$S_x \begin{pmatrix} 1 \\ -1 \end{pmatrix} = -\frac{\hbar}{2} \begin{pmatrix} 1 \\ -1 \end{pmatrix} \quad . \tag{3.23}$$

One may then expand the eigenvector belonging to $S_z' = \frac{1}{2}\hbar$ in terms of the two eigenvectors of S_x:

$$\begin{pmatrix} 1 \\ 0 \end{pmatrix} = \frac{1}{\sqrt{2}} \begin{pmatrix} 1/\sqrt{2} \\ 1/\sqrt{2} \end{pmatrix} + \frac{1}{\sqrt{2}} \begin{pmatrix} 1/\sqrt{2} \\ -1/\sqrt{2} \end{pmatrix} \tag{3.24}$$

or

$$\left| S_z' = \frac{\hbar}{2} \right\rangle = \frac{1}{\sqrt{2}} \left| S_x' = \frac{\hbar}{2} \right\rangle + \frac{1}{\sqrt{2}} \left| S_x' = -\frac{\hbar}{2} \right\rangle \tag{3.24a}$$

Therefore, if an atom is prepared in the state $S_z' = \frac{\hbar}{2}$, and if a second experiment is then made to detect $S_x' = \frac{\hbar}{2}$ only, the probability amplitude will be $\frac{1}{\sqrt{2}}$ or the probability will be

$$P\left(S_x' = \frac{\hbar}{2}\right) = \frac{1}{2} \quad . \tag{3.25}$$

The transformation from the S_z basis to the S_x basis is unitary.

2.4 OBSERVABLES WITH CONTINUOUS SPECTRA

Unlike the spin, the position $(\underset{\sim}{x})$ and the momentum $(\underset{\sim}{p})$ have continuous spectra.[5] They may also be discussed by the

Dirac method, however, provided that one is prepared to introduce continuous matrices and the δ-function. This procedure then treats all observables in the same formal way. Those with discrete spectra are represented by matrices with discrete indices; while others with continuous spectra are represented by matrices that have continuous indices.

The Dirac method, like other methods for demonstrating the equivalence of matrix and wave mechanics, depends on the correspondence between vector and function spaces and on the association of matrices with differential and integral linear operators. There is an elementary discussion of this association in Appendix B. When applied to an operator like the angular momentum with a discrete spectrum there is no difficulty; but observables with continuous spectra require more powerful methods. Nevertheless we shall adopt the Dirac approach and assume that all physical operators may be represented by matrices.[6] This approach depends on the introduction of the Dirac δ-function which is the generalization of the Kronecker δ-function to continuous indices. The Kronecker δ-function satisfies the relation:

$$f(n) = \sum_m \delta(n,m)\, f(m) \quad .$$

The required generalization is

$$f(x) = \int \delta(x-y)\, f(y)\, dy$$

where $\delta(x-y)$ is the Dirac δ-function.

(a) The x-representation

Let the eigenstates and eigenvalues of $\mathbf{x}$ be $|\mathbf{x}'\rangle$ and $\mathbf{x}'$. Then

$$\mathbf{x}|\mathbf{x}'\rangle = \mathbf{x}'|\mathbf{x}'\rangle \tag{4.1}$$

and the expansion of an arbitrary state $|\ \rangle$ is

$$|\ \rangle = \int |\mathbf{x}'\rangle\, d\mathbf{x}'\, \langle\mathbf{x}'|\ \rangle \quad . \tag{4.2}$$

The orthogonality relations now read

$$\langle\mathbf{x}'|\mathbf{x}''\rangle = \delta(\mathbf{x}' - \mathbf{x}'')$$

where $\delta(\mathbf{x}'-\mathbf{x}'')$ is the continuous δ-function,[6] and therefore

$$\begin{aligned}\langle\mathbf{x}'|\mathbf{x}|\mathbf{x}''\rangle &= \mathbf{x}''\, \langle\mathbf{x}'|\mathbf{x}''\rangle \\ &= \mathbf{x}''\, \delta(\mathbf{x}' - \mathbf{x}'') \quad .\end{aligned}$$

Define the wave function

$$\psi(\mathbf{x}') = \langle\mathbf{x}'|\ \rangle \quad . \tag{4.3}$$

Then

$$P(\mathbf{x}') = |\langle\mathbf{x}'|\rangle|^2 = |\psi(\mathbf{x}')|^2 \quad . \tag{4.4}$$

The normalization condition is now

$$\int |\langle \underset{\sim}{x}'|\rangle|^2 \, d\underset{\sim}{x}' = \int |\psi(\underset{\sim}{x}')|^2 \, d\underset{\sim}{x}' = 1 \quad . \tag{4.5}$$

Next consider the operator D_s such that

$$D_s|\rangle = \int |x'\rangle \frac{\partial}{\partial x_s'} \langle x'|\rangle \, dx' \quad .$$

Then

$$x_\ell D_s|\rangle = \int x_\ell' |\underset{\sim}{x}'\rangle \frac{\partial}{\partial x_s'} \langle \underset{\sim}{x}'|\rangle \, d\underset{\sim}{x}'$$

$$= \int |\underset{\sim}{x}'\rangle x_\ell' \frac{\partial}{\partial x_s'} \langle \underset{\sim}{x}'|\rangle \, d\underset{\sim}{x}'$$

Similarly

$$x_\ell|\rangle = \int |\underset{\sim}{x}'\rangle \langle \underset{\sim}{x}'|x_\ell|\rangle \, d\underset{\sim}{x}'$$

$$D_s x_\ell|\rangle = \int |\underset{\sim}{x}'\rangle \frac{\partial}{\partial x_s'} \langle \underset{\sim}{x}'|x_\ell|\rangle \, d\underset{\sim}{x}'$$

$$= \int |\underset{\sim}{x}'\rangle \frac{\partial}{\partial x_s'} x_\ell' \langle \underset{\sim}{x}'|\rangle \, d\underset{\sim}{x}'$$

$$(x_\ell D_s - D_s x_\ell)|\rangle = \int |\underset{\sim}{x}'\rangle \left(x_\ell' \frac{\partial}{\partial x_s'} - \frac{\partial}{\partial x_s'} x_\ell' \right) \langle \underset{\sim}{x}'|\rangle \, d\underset{\sim}{x}'$$

$$= -\delta_{s\ell} \int |\underset{\sim}{x}'\rangle \langle \underset{\sim}{x}'|\rangle \, d\underset{\sim}{x}' \tag{4.6}$$

or

$$(x_\ell D_s - D_s x_\ell)| > = - \delta_{s\ell}| > \quad . \tag{4.7}$$

Then if

$$p_s = \frac{\hbar}{i} D_s \tag{4.8}$$

one obtains the commutator:

$$(x_\ell, p_s)| > = i\hbar\, \delta_{\ell s}| > \tag{4.9}$$

where $| >$ is an arbitrary state.

Therefore we may choose to represent the momentum operator as follows:

$$p_s| > = \int |\underset{\sim}{x}'> \frac{\hbar}{i} \frac{\partial}{\partial x'_s} <\underset{\sim}{x}'| > d\underset{\sim}{x}' \tag{4.10}$$

in order to satisfy the commutation relations (1.11) and (1.12) of this chapter. Then

$$<\underset{\sim}{x}''|p_s| > = \int <\underset{\sim}{x}''|\underset{\sim}{x}'> \frac{\hbar}{i} \frac{\partial}{\partial x'_s} <\underset{\sim}{x}'| > d\underset{\sim}{x}'$$

$$= \frac{\hbar}{i} \frac{\partial}{\partial x''_s} <\underset{\sim}{x}''| > \quad , \tag{4.11}$$

$$<x''|p_s|x'> = \frac{\hbar}{i} \frac{\partial}{\partial x''_s} <x''|x'> \tag{4.12}$$

$$= \frac{\hbar}{i} \frac{\partial}{\partial x''_s} \delta(x'' - x') \quad . \tag{4.13}$$

(b) The Momentum Representation

The basis functions are now eigenstates of momentum. Then

$$p_s|\underset{\sim}{p}'\rangle = p_s'|\underset{\sim}{p}'\rangle \tag{4.14}$$

and

$$|\ \rangle = \int |\underset{\sim}{p}'\rangle\, d\underset{\sim}{p}'\, \langle\underset{\sim}{p}'|\ \rangle \quad . \tag{4.15}$$

Define the momentum amplitude

$$\phi(\underset{\sim}{p}') = \langle\underset{\sim}{p}'|\ \rangle \quad . \tag{4.16}$$

Then

$$P(\underset{\sim}{p}') = |\langle\underset{\sim}{p}'|\rangle|^2 = |\phi(\underset{\sim}{p}')|^2 \tag{4.17}$$

and the normalization is

$$\int P(\underset{\sim}{p}')\, d\underset{\sim}{p}' = \int |\phi(\underset{\sim}{p}')|^2\, d\underset{\sim}{p}' = 1 \quad . \tag{4.18}$$

(c) The Transformation Function $\langle\underset{\sim}{x}'|\underset{\sim}{p}'\rangle$

By (4.2)

$$|\underset{\sim}{p}'\rangle = \int |\underset{\sim}{x}'\rangle\, d\underset{\sim}{x}'\, \langle\underset{\sim}{x}'|\underset{\sim}{p}'\rangle \tag{4.19}$$

and therefore by (4.10)

$$p_s|\underset{\sim}{p}'\rangle = \int |\underset{\sim}{x}'\rangle\, d\underset{\sim}{x}'\, \frac{\hbar}{i}\frac{\partial}{\partial x_s'}\langle\underset{\sim}{x}'|\underset{\sim}{p}'\rangle \tag{4.20}$$

as well as

$$p_s|\underset{\sim}{p}'\rangle = p'_s|\underset{\sim}{p}'\rangle$$
$$= \int |\underset{\sim}{x}'\rangle\, d\underset{\sim}{x}'\; p'_s \langle\underset{\sim}{x}'|\underset{\sim}{p}'\rangle \quad . \tag{4.21}$$

By comparing (4.20) and (4.21) one obtains

$$\frac{\hbar}{i}\frac{\partial}{\partial x'_s}\langle\underset{\sim}{x}'|\underset{\sim}{p}'\rangle = p'_s\langle\underset{\sim}{x}'|\underset{\sim}{p}'\rangle \quad . \tag{4.22}$$

The solution of this first order equation is

$$\langle\underset{\sim}{x}'|\underset{\sim}{p}'\rangle = A\exp\left(\frac{i}{\hbar}\underset{\sim}{x}'\underset{\sim}{p}'\right) \quad . \tag{4.23}$$

The normalization constant may be determined from

$$\langle\underset{\sim}{x}'|\underset{\sim}{x}''\rangle = \int \langle\underset{\sim}{x}'|\underset{\sim}{p}'\rangle\langle\underset{\sim}{p}'|\underset{\sim}{x}''\rangle\, d\underset{\sim}{p}'$$
$$= A^2\int \exp\left(\frac{i}{\hbar}(\underset{\sim}{x}'-\underset{\sim}{x}'')\underset{\sim}{p}'\right) d\underset{\sim}{p}' = \delta(\underset{\sim}{x}'-\underset{\sim}{x}'') \quad .$$

Therefore (see exercise 5 at end of this chapter)

$$A = \left(\frac{1}{2\pi\hbar}\right)^{3/2} \quad . \tag{4.24}$$

The transformation from the $|\underset{\sim}{x}'\rangle$ to the $|\underset{\sim}{p}'\rangle$ basis is a Fourier transformation:

$$\langle\underset{\sim}{p}'|\;\rangle = \int \langle\underset{\sim}{p}'|\underset{\sim}{x}'\rangle\langle\underset{\sim}{x}'|\;\rangle\, d\underset{\sim}{x}' \tag{4.25}$$

or

$$\phi(\underset{\sim}{p}') = \frac{1}{(2\pi\hbar)^{3/2}} \int \exp\left(-\frac{i}{\hbar}\underset{\sim}{x}'\underset{\sim}{p}'\right) \psi(\underset{\sim}{x}')\, d\underset{\sim}{x}' \tag{4.26}$$

$$\psi(\underset{\sim}{x}') = \frac{1}{(2\pi\hbar)^{3/2}} \int \exp\left(\frac{i}{\hbar}\underset{\sim}{x}'\underset{\sim}{p}'\right) \phi(\underset{\sim}{p}')\, d\underset{\sim}{p}' \quad . \tag{4.27}$$

(d) Uncertainty Principle

The probability of $\underset{\sim}{x}'$ for the given state is $|\psi(\underset{\sim}{x}')|^2$ while the probability of $\underset{\sim}{p}'$ for the same state is $|\phi(\underset{\sim}{p}')|^2$. Since these are related by a Fourier transformation, a sharp distribution in $\underset{\sim}{x}$ (or $\underset{\sim}{p}$) space corresponds to a wide distribution in p (or x) space. The product of the dispersions is never less than a minimum value:

$$<(\Delta x)^2> \; <(\Delta p)^2> \; \geq \frac{\hbar^2}{4} \quad . \tag{4.28}$$

This is the uncertainty principle.[7]

(e) Analogue with Classical Contact Transformations

In the $|\underset{\sim}{x}'>$ basis one has

$$<\underset{\sim}{x}''|x_k|\underset{\sim}{x}'> = x_k' \,\delta(\underset{\sim}{x}'' - \underset{\sim}{x}') \tag{4.29a}$$

$$<\underset{\sim}{x}''|p_k|\underset{\sim}{x}'> = \frac{\hbar}{i}\frac{\partial}{\partial x_k''}\,\delta(\underset{\sim}{x}'' - \underset{\sim}{x}') \quad . \tag{4.29b}$$

Similarly in the $|\underset{\sim}{p}'>$ basis one has

$$<\underset{\sim}{p}''|x_k|\underset{\sim}{p}'> = -\frac{\hbar}{i}\frac{\partial}{\partial p_k''}\delta(\underset{\sim}{p}'' - \underset{\sim}{p}') \qquad (4.30a)$$

$$<\underset{\sim}{p}''|p_k|\underset{\sim}{p}'> = p_k''\,\delta(\underset{\sim}{p}'' - \underset{\sim}{p}') \quad . \qquad (4.30b)$$

To obtain (4.30) from (4.29) make the substitution $\underset{\sim}{x} \to \underset{\sim}{p}$ and $\underset{\sim}{p} \to -\underset{\sim}{x}$ as in (2.13) of Chapter 1. This interchange of $\underset{\sim}{p}$ and $\underset{\sim}{x}$ is the analogue of the classical contact transformation (2.13) of Chapter 1.

(f) Unitarity

The Fourier transformation preserves the scalar product

$$\int |\phi(\underset{\sim}{p})|^2\, d\underset{\sim}{p} = \int |\psi(\underset{\sim}{x})|^2\, d\underset{\sim}{x} = 1 \qquad (4.31)$$

and is therefore unitary. The eigenfunctions of the integral transformation (4.27) are the hermite functions.[8] That is

$$\sqrt{2\pi}\;\phi_n(x') = \lambda_n \int_{-\infty}^{+\infty} e^{ix'p'}\;\phi_n(p')\, dp' \qquad (4.32)$$

where

$$\begin{aligned}\phi_n(p) &= (-)^n \exp\left(\frac{1}{2}\,p^2\right)\left(\frac{d}{dp}\right)^n \exp(-p^2)\\ &= \exp\left(-\frac{1}{2}\,p^2\right) H_n(p) \quad .\end{aligned} \qquad (4.33)$$

The eigenvalues are

$$\lambda_n = i^{-n} \quad . \tag{4.34}$$

These are of absolute value unity since the transformation is unitary.

2.5 CHANGE OF BASIS AND UNITARY TRANSFORMATIONS

It has already been remarked that the change from one orthonormal basis to another is unitary. The change from the S_z to the S_x basis is one example, and from the x to the p-basis in (4.32) is another. Let us now consider such transformations in more detail.

The physical interpretation depends on equations like

$$A|a\rangle = a|a\rangle \quad . \tag{5.1}$$

This equation states that a measurement of A is certain to lead to the result a if the system is in the eigenstate $|a\rangle$. To write the same equation in a different reference system, multiply on the left by U. Then

$$UA|a\rangle = aU|a\rangle$$

and

$$(UAU^{-1})\ (U|a\rangle) = a(U|a\rangle) \quad .$$

Let

$$|a\rangle' = U|a\rangle \tag{5.2}$$

$$A' = UAU^{-1} \quad . \tag{5.3}$$

Then

$$A'|a\rangle' = a|a\rangle' \tag{5.4}$$

and according to (5.4) a measurement of the same observable (A') on the same state ($|a\rangle'$) in the primed representation leads to the same result (a) as in the unprimed system.

We next consider equations of motion. These are invariant under a constant similarity transformation of all matrices, namely:

$$A' = UAU^{-1}$$
$$H' = UHU^{-1} \quad . \tag{5.5}$$

For then

$$\frac{dA}{dt} = \frac{(A, H)}{i\hbar} \tag{5.6}$$

$$U \frac{dA}{dt} U^{-1} = U \frac{(A, H)}{i\hbar} U^{-1}$$

$$\frac{dA'}{dt} = \frac{(A', H')}{i\hbar} \quad . \tag{5.7}$$

In equations (5.3) and (5.5) we have not restricted U;

it must be unitary, however, in order to guarantee $\langle a|a\rangle = \langle a|a\rangle'$, or the correct normalization of probability.[9]

It follows that the equations of matrix mechanics are invariant under the group of constant unitary transformations. The actual matrix entries may be changed by such transformations, however, and in this way there arises the arbitrariness mentioned before in 2.2.

If U depends on the time, then

$$\frac{dA'}{dt} = \left(\frac{dU}{dt} U^{-1}\right) A' + U \frac{dA}{dt} U^{-1} + A'U \frac{dU^{-1}}{dt}$$

but

$$\frac{dU}{dt} U^{-1} + U \frac{d}{dt} U^{-1} = 0 \qquad .$$

Let

$$\frac{1}{i\hbar} G = \frac{dU}{dt} U^{-1} \qquad . \tag{5.8}$$

Then

$$\frac{dA'}{dt} = \frac{1}{i\hbar} GA' + U \frac{dA}{dt} U^{-1} + A' \left(- \frac{G}{i\hbar}\right)$$

$$= \frac{1}{i\hbar} (G, A') + U \frac{(A, H)}{i\hbar} U^{-1}$$

$$i\hbar \frac{dA'}{dt} = (G, A') + (A', H')$$

$$= (A', H' - G)$$

$$= (A', K) \qquad (5.9)$$

where

$$K = H' - i\hbar \frac{dU}{dt} U^{-1} \quad . \qquad (5.10)$$

Finally if we substitute

$$\frac{dF}{dt} = - i\hbar \frac{dU}{dt} U^{-1} \qquad (5.11)$$

in (5.10), then

$$K = H' + \frac{dF}{dt} = UHU^{-1} + \frac{dF}{dt} \quad . \qquad (5.12)$$

Therefore (5.6) becomes (5.9) where the new Hamiltonian K is given by either (5.10) or (5.12).

Therefore the equations of motion still remain invariant under unitary transformations. If the unitary transformation depends on the time, however, then the new Hamiltonian is given by either (5.10) or (5.12).

Since unitary transformations preserve the (Heisenberg) equations of motion they corresponds to contact transformations of classical theory which preserve the (Hamiltonian) equations of motion. Similarly unitary transformations preserve the canonical commutators like (q_k, p_ℓ) just as contact transformations preserve the corresponding Poisson brackets.

Infinitesimal Unitary Transformations

A unitary transformation may always be written in the form

$$U = \exp\left(\frac{i}{\hbar} \lambda G\right) \tag{5.13}$$

where G is hermitian and λ is real. Then if λ is infinitesimal, say $\delta\alpha$,

$$U = 1 + \frac{i}{\hbar} \delta\alpha G \tag{5.14}$$

and

$$\begin{aligned} A' &= UAU^{-1} \\ &= \left(1 + \frac{i}{\hbar} \delta\alpha G\right) A \left(1 - \frac{i}{\hbar} \delta\alpha G\right) \\ &= A + \frac{\delta\alpha}{i\hbar} (A, G) \end{aligned} \tag{5.15a}$$

or

$$\frac{\delta A}{\delta\alpha} = \frac{(A, G)}{i\hbar} \quad . \tag{5.15b}$$

G is called the generator of the unitary transformation.

The original transformation (5.13) may be obtained by iteration of the infinitesimal transformation (5.14). There is a corresponding commutator expansion which replaces the ordinary Taylor expansion, namely,

$$e^{iG} A e^{-iG} = A + i(G,A) + \frac{i^2}{2!} (G,(G,A)) + \dots \quad (5.16)$$

This is easily checked by choosing a basis in which G is diagonal. Then on the right side

$$\langle G'|(G,A)|G''\rangle = (G' - G'') \langle G'|A|G''\rangle$$

$$\langle G'|(G,(G,A))|G''\rangle = (G' - G'')^2 \langle G'|A|G''\rangle \quad .$$

On the left side

$$\langle G'|e^{iG} A e^{-iG}|G''\rangle = e^{i(G'-G'')} \langle G'|A|G''\rangle \quad .$$

In the G'-basis we then have from (5.16)

$$e^{i(G'-G'')} \langle G'|A|G''\rangle = [1 + i(G'-G'') + \frac{i^2}{2!} (G'-G'')^2 + \dots] \times \langle G'|A|G''\rangle \quad .$$

This last equation is obviously correct. But it is enough to check (5.16) in any particular basis. Therefore (5.16) is proved.

2.6 SCHRÖDINGER PICTURE[10]

By choosing unitary transformations with constant generators one may find equivalent ways of writing the Heisenberg picture [matrix mechanics in the form (5.6)]. In particular, the symmetry transformations do not lead out of the Heisenberg representation.

Next consider a time-dependent generator which is so chosen that

$$K = UHU^{-1} + \frac{dF}{dt} = UHU^{-1} - i\hbar \frac{dU}{dt} U^{-1} = 0 \quad . \tag{6.1}$$

Then

$$\frac{dA'}{dt} = \frac{1}{i\hbar} (A', K) = 0 \quad .$$

The representation obtained in this way is called the Schrödinger picture. Let us put

$$\begin{aligned} H' &= H_s = U H_h U^{-1} \\ A' &= A_s = U A_h U^{-1} \end{aligned} \tag{6.2}$$

where h and s stand for the Heisenberg and Schrödinger representations. Then

$$H_s = i\hbar \frac{dU}{dt} U^{-1}$$

or

$$H_s U = i\hbar \frac{dU}{dt} \quad . \tag{6.3}$$

So far no reference to the equations of motion of the state function has been made and therefore the formalism is actually incomplete. Let $| >_h$ be a state vector in the Heisenberg representation. Then let us complete the definition of the Heisenberg representation by postulating:

$$\frac{d}{dt} | >_h = 0 \quad . \tag{6.4}$$

It now follows from (6.2) and section (2.5) that

$$| >_s = U | >_h \quad . \tag{6.5}$$

Therefore equation (6.3), namely:

$$H_s U | >_h = i\hbar \frac{dU}{dt} | >_h$$

becomes

$$H_s (U | >_h) = i\hbar \frac{d}{dt} (U | >_h)$$

or

$$H_s | >_s = i\hbar \frac{d}{dt} | >_s \quad . \tag{6.6}$$

The equations of the Heisenberg and Schrödinger representations may be summarized as follows:

$$\frac{d}{dt} | \rangle_h = 0$$

$$\frac{dA_h}{dt} = \frac{(A_h, H_h)}{i\hbar} \tag{6.7}$$

and

$$-\frac{\hbar}{i}\frac{d}{dt} | \rangle_s = H_s | \rangle_s$$

$$\frac{dA_s}{dt} = 0 \quad . \tag{6.8}$$

Finally

$$H_h = U^{-1} H_s U$$

$$= H_s \tag{6.9}$$

where

$$U = \exp\left(\frac{1}{i\hbar} H_s t\right) \tag{6.10}$$

according to (6.3).

Therefore in the Heisenberg picture, operators depend on the time but the states are constant, while in the Schrödinger picture the operators are constant and the states depend on the time. The two pictures are so related, however, that matrix <u>elements</u> do not depend on the representation in which they are computed.[11]

Schrödinger Equation

The familiar form of the Schrödinger differential equation may be obtained from (6.6). Drop the subscript s and write

$$H \mid \; > \; = i\hbar \frac{d}{dt} \mid \; > \tag{6.11}$$

where H is, for example,

$$H = \sum \frac{1}{2m_k} \underset{\sim}{p}_k{}^2 + \sum V_{ij}(|\underset{\sim}{x}_i - \underset{\sim}{x}_j|) \quad .$$

Then by introducing the x-representation of $|\;>$ we find

$$\begin{aligned} H(\underset{\sim}{x}_1 \ldots \underset{\sim}{p}_1 \ldots) | \; > \; &= \int \ldots \int \; |\underset{\sim}{x}_1{}' \ldots > \; H(\underset{\sim}{x}_1{}' \ldots \frac{\hbar}{i} \frac{\partial}{\partial \underset{\sim}{x}_1'} \ldots) \\ &\qquad \times <\underset{\sim}{x}_1{}' \ldots | \; > \; d\underset{\sim}{x}_1{}' \ldots \\ &= \int \ldots \int \; |\underset{\sim}{x}_1{}' \ldots > \; H(\underset{\sim}{x}_1{}' \ldots \frac{\hbar}{i} \frac{\partial}{\partial \underset{\sim}{x}_1'} \ldots) \\ &\qquad \times \psi(\underset{\sim}{x}_1{}' \ldots t) \; d\underset{\sim}{x}_1{}' \ldots \end{aligned} \tag{6.12}$$

and by (6.11)

$$= i\hbar \int \ldots \int \; |\underset{\sim}{x}_1{}' \ldots > \frac{\partial \psi}{\partial t} (\underset{\sim}{x}_1{}' \ldots t) \; d\underset{\sim}{x}_1{}' \ldots . \tag{6.13}$$

By comparing (6.12) and (6.13) one finds

$$H(\underset{\sim}{x}_1{}'\ldots \frac{\hbar}{i}\frac{\partial}{\partial \underset{\sim}{x}_1{}'}\ldots)\,\psi(\underset{\sim}{x}_1{}'\ldots t) = i\hbar\frac{\partial}{\partial t}\psi(\underset{\sim}{x}_1{}'\ldots t) \quad . \tag{6.14}$$

If the N-particle gas is composed of identical particles, then the wave function is subject to an additional constraint: it is either symmetric or antisymmetric in the exchange of coordinates of any pair of particles. If such a supplementary condition is imposed as a boundary condition at one time, it will remain true for all time.[12]

In the Schrödinger picture one may associate a wave packet with the representative point $(\underset{\sim}{x}_1\ldots\underset{\sim}{x}_N)$ in a 3N dimensional configuration space, and for the case of weakly interacting gases one may assign wave packets to the individual particles in three dimensional space. In the classical limit, say for planetary systems, the size of the packets is small compared to the radius of the orbits. The wave packet description thus provides a very intuitive way of passing to the classical limit.

2.7 INVARIANCE GROUP AND LIE ALGEBRA OF NON-RELATIVISTIC QUANTUM THEORY

When its generator is independent of the time, a unitary transformation may be said to belong to the invariance group. Let these generators be denoted by X_i.

Then

$$\frac{dX_i}{dt} = 0 \quad . \tag{7.1}$$

If ξ_i is the parameter corresponding to X_i, then by (5.15)

$$\frac{dA}{d\xi_i} = \frac{(A,\ X_i)}{i\hbar} \quad . \tag{7.2}$$

By (7.1) and (5.12), when U_i does not depend on the time,

$$K = H' = U_i H U_i^{-1} \quad , \tag{7.3}$$

where U_i is generated by X_i. Finally the X_i do not in general commute with each other but belong to a Lie Algebra

$$(X_m,\ X_n) = i\hbar\ C_{mn}{}^p\ X_p \tag{7.4}$$

where the $C_{mn}{}^p$ are called the structure constants.

The systems considered in this book are characterized by an invariance group with the generators $(\underset{\sim}{P},\ \underset{\sim}{L},\ \underset{\sim}{G},\ H)$.

Then (7.2) becomes

$$\frac{dA}{dt} = \frac{1}{i\hbar}\ (A,\ H)$$

$$\frac{dA}{dx} = \frac{1}{i\hbar}\ (A,\ P_x)$$

$$\frac{dA}{d\omega_{xy}} = \frac{1}{i\hbar}(A, L_{xy})$$

$$\frac{dA}{dV_x} = \frac{1}{i\hbar}(A, G_x) \quad . \tag{7.5}$$

It was shown earlier that the equations of quantum theory are invariant under constant unitary transformations. One now sees that these equations themselves describe infinitesimal unitary transformations.

The equations (7.5) have the solution

$$A(\xi) = U_\xi \, A(0) \, U_\xi^{-1} \tag{7.6}$$

where U_ξ in the different cases stands for the following:

$$U_t = \exp\left(-\frac{1}{i\hbar} Ht\right) \tag{7.6a}$$

$$U_x = \exp\left(-\frac{1}{i\hbar} P_x x\right) \tag{7.6b}$$

$$U_{xy} = \exp\left(-\frac{1}{i\hbar} L_{xy} \, \omega_{xy}\right) \tag{7.6c}$$

$$U_v = \exp\left(-\frac{1}{i\hbar} G_x V_x\right) \quad . \tag{7.6d}$$

Equation (7.6) means, for example, when $\xi = t$,

$$A(t) = \exp\left(-\frac{1}{i\hbar} Ht\right) A(0) \exp\left(\frac{1}{i\hbar} Ht\right) \quad . \tag{7.7}$$

Multiplying on the left and right with eigenstates of

H one obtains

$$\langle n|A(t)|m\rangle = \langle n|A(0)|m\rangle \exp\left(\frac{1}{i\hbar}(E_m - E_n)t\right) \quad .$$

These U are called displacement operators. They comprise, except for the time displacement itself, a group of time independent unitary operators. They provide a representation of the symmetry group while the hermitian exponents provide a representation of the corresponding Lie algebra.

One sees that the algebra of the generators is just the same in the quantum theory as in the classical limit of the quantum theory. In the one case the algebra is realized by the commutators and in the other by the Poisson brackets. The various quantum conservation laws follow from (7.5) just as the corresponding classical statements follow from the Poisson bracket relations. (See paragraph 4 of Chapter 1.)

2.8 QUANTUM NUMBERS AND S-MATRIX

(a) Complete Set of Quantum Numbers

In general, one may construct from the complete algebra of constant generators (X_i) a smaller set of objects (Λ_i) which commute with each other. (The general construction is given in Appendix D.) Then

$$\frac{d\Lambda_i}{dt} = 0 \tag{8.1}$$

and

$$(\Lambda_i, \Lambda_j) = 0 \tag{8.2}$$

while

$$\frac{dX_i}{dt} = 0$$

but

$$(X_i, X_j) = i\hbar\, C_{ij}{}^m X_m \neq 0 \qquad \text{in general.}$$

Since the Λ_i commute, they can be diagonalized simultaneously and their simultaneous eigenvalues may be used to label the quantum states: $|\lambda_1{}' \lambda_2{}' \cdots\rangle$. By definition $(\lambda_1{}' \lambda_2{}' \cdots)$ are the quantum numbers. The Λ_i include the so-called Casimir operators C_α which have the property of commuting with the complete algebra:

$$(C_\alpha, X_i) = 0 \quad . \tag{8.3}$$

The number of independent C_α is known as the rank of the group.

For example, from the angular momentum operators (L_x, L_y, L_z) one may construct (L^2, L_z) which commute with each other as well as with the Hamiltonian. In this case L^2 is the Casimir operator.

(b) S-Matrix

By (3.16) the conservation of probability requires that the norm of a physical state be time independent, and the solution of the Schrödinger equation may therefore be expressed as follows:

$$|t''\rangle = U(t'',t')\,|t'\rangle \tag{8.4}$$

where U is a unitary operator. For example by (6.10)

$$U(t'', t') = \exp\left(\frac{1}{i\hbar} H(t'' - t')\right) \quad . \tag{8.5}$$

The corresponding Heisenberg equation is (7.7) or

$$A(t) = U_t\, A(0)\, U_t^{-1} \quad . \tag{8.6}$$

Let

$$S = U(+\infty, -\infty) \quad . \tag{8.7}$$

Now let $|a_1' a_2' \ldots\rangle \ldots = |a'\rangle \ldots$ be some complete set of states specified at the beginning of the experiment. These may be called "in-states." Suppose that the system is initially in the particular in-state $|a'\rangle$. Then at $t = +\infty$ this initial state will have evolved into $S|a'\rangle$. The probability amplitude of finding the system at $t = +\infty$ in the different in-state $|a''\rangle$ is $\langle a''|S|a'\rangle$. S is called the S-matrix and is unitary by construction.[13]

Let A be any observable that commutes with S:

$$(A, S) = 0 \quad . \tag{8.8}$$

Take this equation between eigenstates of A. Then

$$\langle a'|(S,A)|a''\rangle = 0$$

$$(a' - a'')\langle a'|S|a''\rangle = 0 \quad . \tag{8.9}$$

Therefore if $a' \neq a''$,

$$\langle a'|S|a''\rangle = 0 \quad , \tag{8.10}$$

i.e., the probability of a transition between two states with different values of a' vanishes, or A is an integral of the motion.

If A is not an integral of the motion, then $\langle a'|S|a''\rangle \neq 0$ and $|\langle a'|S|a''\rangle|^2$ is the probability of a transition for a' to a''.

In summary, constants of the motion (X_i) are distinguished from other observables by the fact that they commute with the S-matrix and with the Hamiltonian. Although they do not in general commute among themselves, they close an algebra, and by exponentiation generate the symmetry group. Finally from the X_i one may form the maximal commuting set $\{\Lambda_i\}$ that provides a complete set of quantum numbers.

2.9 STATISTICAL PHYSICS

(a) Classical Description of Macroscopic State

Since we are interested in a general physical system, which may be of macroscopic size, we cannot in general give an exact description even classically. That is because the given macroscopic state is compatible with very many microscopic states.

To describe the classical system it is useful to introduce phase space, which is the 2f-dimensional space of the generalized coordinates and momenta. The precise state of the dynamical system may be represented by a single point in phase space and the complete dynamical evolution of the system is then represented by the motion of this point in phase space according to Hamilton's equations. A thermodynamic, or a macro-

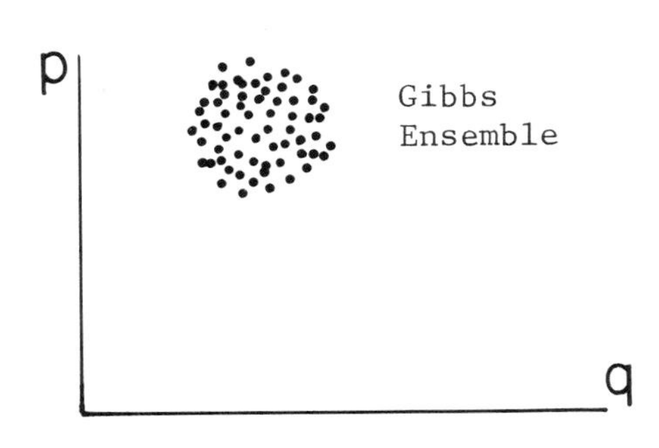

Fig. 2.1 Phase Space

scopic state, is represented by an ensemble of points in phase space. This collection of microscopic states, all compatible with our knowledge of the macroscopic system, is called the Gibbs ensemble.

The expectation value of A for such a macroscopic state is

$$\langle A \rangle = \int A\rho d\tau \tag{9.1}$$

where

$$\int \rho d\tau = 1 \tag{9.2}$$

$$d\tau = dq_1 \ldots dq_f \; dp_1 \ldots dp_f \quad . \tag{9.3}$$

Here ρ is the density of representative points in the Gibbs ensemble. Since each point represents a different copy of the macroscopic system, these points are conserved and ρ obeys the equation of continuity

$$\frac{\partial \rho}{\partial t} + \mathrm{div}\; (\rho \underset{\sim}{v}) = 0 \quad . \tag{9.4}$$

Here $\underset{\sim}{v}$ is the velocity:

$$\underset{\sim}{v} = \left(\frac{dq_1}{dt} \ldots \frac{dq_f}{dt} \; \frac{dp_1}{dt} \ldots \frac{dp_f}{dt} \right) \quad . \tag{9.5}$$

By (9.4)

$$\frac{\partial \rho}{\partial t} + (\vec{v} \cdot \text{grad})\rho + \rho \text{ div } \vec{v} = 0$$

or

$$\frac{D\rho}{Dt} + \rho \text{ div } \vec{v} = 0 \tag{9.6}$$

where

$$\frac{D\rho}{Dt} = \frac{\partial \rho}{\partial t} + (\vec{v} \cdot \text{grad})\rho \tag{9.7}$$

is the hydrodynamical derivative.

By Hamilton's equations

$$\text{div } \vec{v} = \frac{\partial}{\partial q_1} \frac{dq_1}{dt} + \ldots + \frac{\partial}{\partial p_1} \frac{dp_1}{dt} + \ldots$$

$$= \left(\frac{\partial^2 H}{\partial q_1 \partial p_1} + \ldots\right) - \left(\frac{\partial^2 H}{\partial p_1 \partial q_1} + \ldots\right)$$

or

$$\text{div } \underset{\sim}{v} = 0 \quad . \tag{9.8}$$

It follows that

$$\frac{D\rho}{Dt} = 0 \quad . \tag{9.9}$$

This is the Liouville theorem. It may be written in P.B.

form as follows:

$$\frac{D\rho}{Dt} = \frac{\partial\rho}{\partial t} + \sum_k \left(\frac{dq_k}{dt}\frac{\partial}{\partial q_k} + \frac{dp_k}{dt}\frac{\partial}{\partial p_k}\right)\rho = 0$$

$$= \frac{\partial\rho}{\partial t} + \sum_k \left(\frac{\partial H}{\partial p_k}\frac{\partial}{\partial q_k} - \frac{\partial H}{\partial q_k}\frac{\partial}{\partial p_k}\right)\rho = 0$$

or

$$\frac{\partial\rho}{\partial t} = -\ [\rho,\ H] \quad . \tag{9.10}$$

(b) Quantum Description of Macroscopic State

An actual physical system of macroscopic size is almost always represented by an incoherent mixture of several quantum states. Such an incoherent mixture is recognized as a macroscopic state and is represented by an ensemble in <u>state</u> space. If the system is in thermodynamic equilibrium, then this macroscopic state is a thermodynamic state.

Near absolute zero a macroscopic system is described by a very small number of quantum states (third law). Because of the resulting large amount of coherence the behavior of such a system is very unclassical (e.g.,

Fig. 2.2 State Space

superfluids). The expectation value of A is in general

$$\langle A\rangle = \sum_{\alpha} W_{\alpha} \langle\alpha|A|\alpha\rangle \tag{9.11}$$

with the normalization

$$\sum_{\alpha} W_{\alpha} = 1 \quad . \tag{9.12}$$

Here the sum is over the quantum states (α) which are mixed incoherently with weights W_{α}. The mixing is incoherent because the averages are taken before the sum over the ensemble.

It follows from (9.11)

$$\langle A\rangle = \sum_{s,t,\alpha} W_{\alpha} \langle\alpha|s\rangle \langle s|A|t\rangle \langle t|\alpha\rangle \quad . \tag{9.13}$$

Define the density matrix ρ by

$$\langle t|\rho|s\rangle = \sum_{\alpha} \langle t|\alpha\rangle W_{\alpha} \langle\alpha|s\rangle \quad . \tag{9.14}$$

Then from (9.12) and (9.13)

$$\langle A\rangle = \sum_{s,t} \langle s|A|t\rangle \langle t|\rho|s\rangle \tag{9.15}$$

or

$$\langle A\rangle = \mathrm{Tr}\, A\rho = \mathrm{Tr}\, \rho A$$

where Tr means trace. Then

$$\begin{aligned} \mathrm{Tr}\ \rho &= \sum_t \langle t|\rho|t\rangle \\ &= \sum_{\alpha t} \langle t|\alpha\rangle\, W_\alpha\, \langle\alpha|t\rangle \\ &= \sum_\alpha W_\alpha = 1 \end{aligned}$$

or

$$\mathrm{Tr}\ \rho = 1 \quad . \tag{9.16}$$

The condition for a pure state is

$$\rho^2 = \rho \quad . \tag{9.17}$$

Finally we give the quantal form of the Liouville theorem. From the definition of the density matrix (9.14) it follows that

$$\rho = \sum_\alpha |\alpha\rangle\, W_\alpha\, \langle\alpha| \quad . \tag{9.18}$$

Then

$$\begin{aligned} i\hbar \frac{\partial\rho}{\partial t} &= \sum_\alpha \left[\left(i\hbar \frac{\partial}{\partial t} |\alpha\rangle\right) W_\alpha \langle\alpha| + |\alpha\rangle\, W_\alpha \left(i\hbar \frac{\partial}{\partial t} \langle\alpha|\right)\right] \\ &= \sum_\alpha \left[H|\alpha\rangle\, W_\alpha\, \langle\alpha| - |\alpha\rangle\, W_\alpha\, \langle\alpha|H\right] \\ &= H\rho - \rho H \end{aligned}$$

or

$$\frac{\partial \rho}{\partial t} = - \frac{(\rho, H)}{i\hbar} \quad . \tag{9.19}$$

This is the quantal form of the Liouville theorem. The classical and the quantal formulas are again connected by the substitution

$$[\ ,\ ] \to \frac{(\ ,\ )}{i\hbar} \quad .$$

NOTES ON CHAPTER 2

1. The entire development of this chapter follows selected parts of Dirac's book, reference 1.
2. The arbitrariness results from the possibility of making unitary transformations in quantum theory, and corresponds to the classical possibility of making canonical transformations and the consequent arbitrariness in the functions $x(t)$ and $p(t)$ that describe the classical orbit.
3. Dirac, op. cit.; Landau (2).
4. Dirac, op. cit.; or exercise 17.
5. It is clear that the canonical commutation rules (1.11) cannot be satisfied by finite matrices. [If one takes the trace of (1.11) one finds for finite matrices the contradiction: $0 = i\hbar N$ where N is the dimensionality

of the matrices.] Therefore relations like (1.11) cannot be satisfied by finite matrices. On the other hand it is possible to satisfy relations like (1.17) in this way, as one knows from the angular momentum algebra.

Notice also that

$$(G_\ell, P_k) = i\hbar M\delta_{\ell k} \tag{N5.1}$$

according to (5.11) of the first chapter. These equations correctly tell us that the Galilean generators also correspond to infinite matrices, since they satisfy commutation rules like (1.11).

It is remarked, however, in note 7, chapter 1, that the above commutation relation becomes

$$(M_{o\ell}, P_k) = i\hbar P_o \delta_{k\ell} \tag{N5.2}$$

in the relativistic theory. Now these commutation rules are like (1.17); nevertheless they still cannot be satisfied by finite matrices. [Since the signature of the Lorentz group is (+++ -), P_o is imaginary if the P_k are real. Then (N5.2) differs from (1.17) by an i on the right and it may again be shown that there are no finite realizations of (N5.2).]

6. After the equivalence of the Heisenberg matrix mechanics and the Schrödinger wave mechanics had been established,

and after the probability interpretation had been discovered by Born, the complete theory was recast by Jordan and by Dirac. As already remarked, the Dirac formalism makes use of continuous matrices and the δ-function and needs to be justified by a more rigorous mathematical analysis. A mathematically satisfactory formulation, growing out of the work of Jordan, was developed by von Neumann[4] who showed that the appropriate mathematical language was the theory of Hilbert spaces.

Although von Neumann avoids the δ-function altogether, there has developed in more recent years a mathematically satisfactory formulation of the δ-function as well. This is the Schwartz theory of distributions. For the δ-function see references (1), (8), and (10), and for the theory of distributions see (10) and references given there.

7. For proof see, for example, Saxon (8).
8. See Titchmarsh (11), page 81.
9. See also Dirac and exercise 16.
10. See Dirac (1).
11. Matrix elements do not depend on the representation. Let

$$|a'\rangle_s = U\ |a'\rangle_h$$

$$A_s = U\ A_h\ U^{-1} \qquad .$$

Then

$$ {}_s\langle a''|A_s|a'\rangle_s = {}_h\langle a''\ U^{-1}(U\ A_h\ U^{-1})U|a'\rangle_h $$

$$ = {}_h\langle a''|A_h|a'\rangle_h \quad . $$

Therefore matrix elements have the same time dependence in the Schrödinger and Heisenberg pictures, and in particular if they are constant in the one representation, they are also constant in the other representation.

12. The wave function is symmetric or antisymmetric according to whether the identical particles have integral or half-integral spin. See exercise 14 in connection with the supplementary condition.

The symmetry constraint is particularly important in systems containing only a few particles, and in atoms it is the origin of the Pauli exclusion principle. However, even in macroscopic systems the quantum entropy is quite different from the classical result because of the symmetry constraint satisfied by the wave function. See, for example, reference 9 at the end of this chapter. Furthermore, at low temperatures macroscopic systems exhibit quantum behavior (superfluidity) and under these conditions also the symmetry of the wave function (the statistics) becomes of the utmost importance.

13. The S-matrix is ordinarily used to connect in- and out-states of a scattering problem. In these problems the in- and out-states describe wave packets that are far apart and either converging at $t = -\infty$ or diverging at $t = +\infty$.

BIBLIOGRAPHY FOR CHAPTER 2

General Works on Quantum Theory

1. P. A. M. Dirac, Quantum Mechanics, Oxford (1958).
2. L. D. Landau and E. M. Lifshitz, Quantum Mechanics, Addison-Wesley (1958).
3. W. Pauli, Die allgemeinen Prinzipien der Wellenmechanik, Encyclopedia of Physics, Vol. 5, Springer, Berlin (1958).
4. J. von Neumann, Mathematical Foundations of Quantum Mechanics, Princeton Univ. Press (1955).
5. L. I. Schiff, Quantum Mechanics, McGraw-Hill (1955).
6. A. Messiah, Quantum Mechanics, North-Holland, Amsterdam (1961).
7. K. Gottfried, Quantum Mechanics, Benjamin (1966).

The following undergraduate text is recommended as an introduction to this chapter.

8. D. S. Saxon, Elementary Quantum Mechanics, Holden-Day (1968).

For quantum mechanics and macroscopic properties see, for example,

9. R. Finkelstein, A Short Introduction to Thermodynamics and Statistical Physics, Freeman (1969).

Mathematical Methods

10. B. Friedman, Principles and Techniques of Applied Mathematics, Wiley (1956).

11. E. C. Titchmarsh, Introduction to the Theory of the Fourier Integral, Oxford (1937).

PROBLEMS

1. Show

$$(A,BC) = (A,B)C + B(A,C)$$

$$(A,(B,C)) + (B,(C,A)) + (C,(A,B)) = 0$$

2. Show

$$\lim_{\hbar\to 0} \frac{(A,B)}{i\hbar} = [A,B]$$

where A and B are polynomials in p and q.

3. Find the state $| \;\rangle$ which describes an electron polarized along $\hat{n}$:

$$(\hat{n}\,\vec{S})|\ \rangle = \frac{\hbar}{2}|\ \rangle$$

Find also

the probability that $S_z = \frac{\hbar}{2}$

the expectation value of S_z: $\langle S_z \rangle$

the expectation value of $(\Delta S_z)^2$: $\langle(\Delta S_z)^2\rangle$,

for the same state.

4. Find

$$e^{\frac{i}{\hbar}P_x a}\ X\ e^{-\frac{i}{\hbar}P_x a}$$

$$e^{\frac{i}{\hbar}L_z\omega}\ X\ e^{-\frac{i}{\hbar}L_z\omega}$$

$$e^{\frac{i}{\hbar}G_x v}\ X\ e^{-\frac{i}{\hbar}G_x v}$$

$$e^{\frac{i}{\hbar}G_x v}\ H\ e^{-\frac{i}{\hbar}G_x v}$$

where P_x and L_z are linear and angular momenta, and where G_x is the generator of a Galilean transformation. Here X is the center of mass.

5. One representation of the Dirac δ-function is

$$\delta(x) = \frac{1}{2\pi}\int_{-\infty}^{+\infty} e^{iax}\,da \quad .$$

Show

$$f(x') = \int_{-\infty}^{+\infty} \delta(x'-x)\ f(x)\ dx \qquad .$$

6. Find the probability $|\phi(p)|^2$ of the particular momentum p for the ground state of hydrogen.

7. Show that the condition for a pure state is $\rho^2 = \rho$.

8. Evaluate

$$\bar{A} = \mathrm{Tr}\ \rho A$$

when $\langle x|A|x'\rangle = A(x)\ \delta(x-x')$. Show that

$$\bar{A} = \sum W_\alpha \int A(x)\ |\psi_\alpha(x)|^2\ dx \qquad .$$

9. Let $H = \frac{p^2}{2m} + V$. By calculating $((H,x),x)$ find sum rule

$$\sum_m |x_{nm}|^2\ (E_m - E_n) = ?$$

10. Given the Hamiltonian

$$H = \frac{p^2}{2m} + V(x)$$

use the Heisenberg representation to show (Ehrenfest's theorem)

$$m\frac{d^2}{dt^2}\langle|\underset{\sim}{x}|\rangle = -\langle|\underset{\sim}{\nabla V}|\rangle$$

for any state $| >$.

11. Show that

$$<|(\Delta A)^2|> <|(\Delta B)^2|> \geq \frac{1}{4} |<|(A,B)|>|^2$$

for any allowed state $| >$.

12. Discuss the spreading of a wave packet. Let

$$\psi(x,0) \sim e^{-x^2/4d^2} .$$

Show that

$$|\psi(x,t)|^2 \sim e^{-x^2/2D^2}$$

where

$$D^2 = d^2 + \hbar^2 t^2/4m^2 d^2 .$$

Does spreading of packet in x-space imply by the uncertainty relation that knowledge of momentum is becoming sharper? One may speak of a closed classical orbit only if the spreading of a wave packet in one period is small compared to the radius of the orbit. Show that the condition for the spreading to be so limited implies that the angular momentum $\gg \hbar$.

Apply similar criterion to decide if classical theory may be used in kinetic theory of gases.

13. Discuss quantum theory and classical limits of the motion of a single particle over and through potential barrier, or well, as illustrated in the following figures

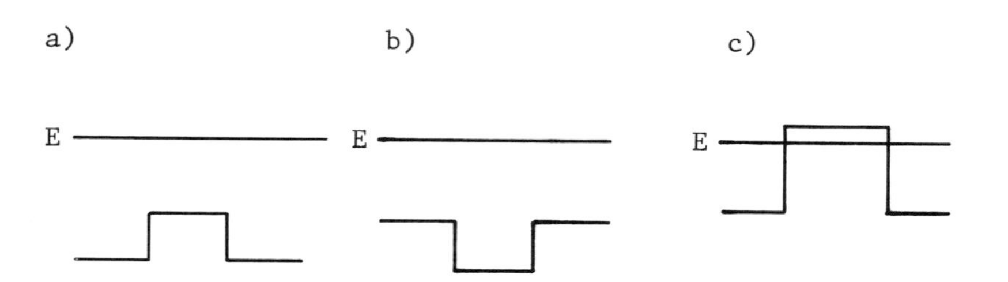

Find transmission coefficients in these three cases. Discuss resonance condition in (a) and (b). Apply result of (c) to discuss α-decay.

For detailed description of motion of packet through barrier see Saxon, page 158, reference 10 of this chapter.

14. By solving the Schrödinger equation, find the eigenfunctions and the eigenvalues of the Hamiltonian of a harmonic oscillator.

15. Since the Hamiltonian of a harmonic oscillator is symmetric in x and p (in suitable units), the eigenstates in the x-representation $\langle x|n\rangle$ and in the p-representation $\langle p|n\rangle$ must be the same. Use (4.32) to find these functions. Then show that

$$\frac{e^{ixp}}{\sqrt{2\pi}} = \sum_n i^n \, \phi_n(x) \, \phi_n(p)$$

or

$$\langle x|p\rangle = \sum_n \langle x|n\rangle \, \langle n|p\rangle$$

where the $\langle x|n\rangle$ are the hermite functions. [Compare with (8.13) of Chapter 3.]

16. Prove that $UU^\dagger = cI$, if every hermitian matrix B remains hermitian under the transformation:

$$A = U \, B \, U^{-1} \quad .$$

17. Show that the transformation from one orthonormal basis $|a'\rangle$ to another $|b'\rangle$ is unitary. Show that

$$\sum_{a'} |\langle a'|\rangle|^2 = \sum_{b'} |\langle b'|\rangle|^2 \quad .$$

18. Show that

$$\begin{aligned}\langle n|p|m\rangle &= \iint \langle n|x'\rangle \, \langle x'|p|x''\rangle \, \langle x''|m\rangle \, dx' \, dx'' \\ &= \int \psi_n(x')^* \, \frac{\hbar}{i} \frac{d}{dx'} \, \psi_m(x') \, dx'\end{aligned}$$

by using (4.13). In order that $\langle n|p|m\rangle$ be a hermitian matrix, how must the class of functions containing $\psi_n(x)$ and $\psi_m(x)$ be restricted?

19. Prove that the permutation operators commute with the Hamiltonian of a set of identical particles. Then show that a symmetric (or antisymmetric) wave function remains symmetric (or antisymmetric).

20. If A is a homogeneous function of degree n in the coordinates, show that

$$U\,A\,U^{-1} = e^{n\phi}\,A$$

where $U = e^{(i/\hbar)\phi D}$, and D is the dilatation operator $\sum x_{i\alpha}\,p_{i\alpha}$. How is U made unitary?

CHAPTER 3

LAGRANGIAN AND VARIATIONAL FORMULATIONS

3.1 HAMILTON'S PRINCIPLE

In the first two chapters we have attempted to make an introductory survey of the principles of classical and quantum mechanics in a rapid and preliminary way. The quantal development has been described in more detail and carried further than the corresponding classical theory, both because the quantum theory provides the underlying description and because it is formally simpler, since its development may be based on algebraic methods. In particular we have discussed the unitary transformations of quantum theory in some detail but have only mentioned the corresponding facts about classical contact transformations. Similarly we have discussed the Schrödinger equation but not the corresponding classical equation, namely the Hamilton-Jacobi equation. At the same time, the quantal development was also incomplete since only the Heisenberg and Schrödinger formulations were described. In the present chapter we shall present the more modern Lagrangian formulations that were pioneered by Dirac and more fully developed by Feynman and Schwinger. Finally, since both the classical and quantal parts were treated in an elementary and not very systematic manner, we shall now start anew in order to present a more orderly and complete development, and we shall begin with classical mechanics.

Let the dynamical system be completely specified by f-data (generalized coordinates): $q_1 \ldots q_f$ and let the motion of this system be represented by a curve in f-dimensional q-space (called configuration space)

$$q_k = q_k(t) \qquad k = 1 \ldots f \tag{1.1}$$

where the parameter, t, is the time.

The actual trajectory leading from (A) to (B) may be selected from all conceivable trajectories by means of a general variational principle (Hamilton's principle) and a particular dynamical function that characterizes the system. This function is the Lagrangian and depends on the coordinates and their first derivatives:

$$L = L(q_1 \ldots q_f, \dot{q}_1 \ldots \dot{q}_f, t) \tag{1.2}$$

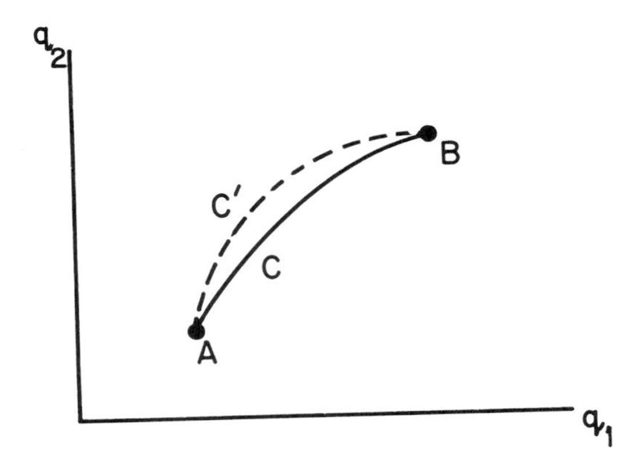

Fig. 1.1 Hamilton's Principle. C and C' are correlated with the same value of the time. $\delta S = 0$ determines AB.

We postulate that the Lagrangian be invariant under the rotation-translation group.

Hamilton's principle states that the following line integral (the action) is an extremal for the actual motion leading from A to B:

$$S = \int_{t_1}^{t_2} L(q_1 \ldots q_f, \dot{q}_1 \ldots \dot{q}_f, t)dt \quad . \tag{1.3}$$

That is

$$\delta S = 0 \quad . \tag{1.4}$$

The procedure implied by (1.4) is that we calculate S for the class of comparison trajectories q(t) which satisfy all the given boundary conditions and constraints. We then select from this class of functions the one which satisfies (1.4).

Let AC'B be one path belonging to this family (a so-called varied path). Let C' be put into correspondence with some point on the actual path C and labeled with the same value of t as C, and let us write

$$q_k'(t) = q_k(t) + \delta q_k(t) \tag{1.5}$$

where $q_k'(t)$ and $q_k(t)$ are the coordinates of C' and C respectively.

Then

$$\frac{dq_k'}{dt} = \frac{dq_k}{dt} + \frac{d}{dt}\,\delta q_k \quad . \tag{1.6}$$

Define

$$\delta\left(\frac{dq_k}{dt}\right) = \left(\frac{dq_k}{dt}\right)' - \left(\frac{dq_k}{dt}\right) = \frac{dq_k'}{dt'} - \frac{dq_k}{dt} \quad . \tag{1.7}$$

By (1.6), since $t' = t$,

$$\delta\left(\frac{dq_k}{dt}\right) = \frac{d}{dt}\,\delta q_k \quad . \tag{1.8}$$

We may now calculate

$$\delta S = \int_{t_1}^{t_2} \sum_k \left(\frac{\partial L}{\partial q_k}\,\delta q_k + \frac{\partial L}{\partial \dot{q}_k}\,\delta \dot{q}_k\right) dt \quad . \tag{1.9}$$

By (1.8)

$$\int_{t_1}^{t_2} \frac{\partial L}{\partial \dot{q}_k} \delta \dot{q}_k \, dt = \int_{t_1}^{t_2} \frac{\partial L}{\partial \dot{q}_k} \left(\frac{d}{dt} \delta q_k \right) dt$$

$$= - \int_{t_1}^{t_2} \left[\frac{d}{dt} \left(\frac{\partial L}{\partial \dot{q}_k} \right) \right] \delta q_k \, dt + $$

$$+ \left[\frac{\partial L}{\partial \dot{q}_k} \delta q_k \right]_{t_1}^{t_2} .$$

Therefore

$$\delta S = \int_{t_1}^{t_2} \sum_k \left[\frac{\partial L}{\partial q_k} - \frac{d}{dt} \left(\frac{\partial L}{\partial \dot{q}_k} \right) \right] \delta q_k(t) \, dt + $$

$$+ \sum_k \left[\frac{\partial L}{\partial \dot{q}_k} \delta q_k \right]_{t_1}^{t_2} . \quad (1.10)$$

To express Hamilton's principle one now limits the class of varied paths by the requirement that they all have the same initial and final points. Then

$$\delta q_k(t_1) = \delta q_k(t_2) = 0 \quad (1.11)$$

and by (1.4)

$$\delta S = \int_{t_1}^{t_2} \sum_k \left[\frac{\partial L}{\partial q_k} - \frac{d}{dt} \left(\frac{\partial L}{\partial \dot{q}_k} \right) \right] \delta q_k(t) \, dt = 0 \ . \qquad (1.12)$$

But the variations $\delta q_k(t)$ are independent and arbitrary. Choose

$$\delta q_\ell(t) = \delta_{k\ell} \, \delta(t-t')\varepsilon \ . \qquad (1.13)$$

where ε is infinitesimal. Then

$$\frac{\partial L}{\partial q_k} - \frac{d}{dt} \left(\frac{\partial L}{\partial \dot{q}_k} \right) = 0 \qquad k = 1 \cdots f \ . \qquad (1.14)$$

Conversely if the set (1.14) is satisfied, then $\delta S = 0$ also. The condition $\delta S = 0$ may not imply a minimum. To decide this question one must determine the kinetic focus.[1]

The equations (1.14) are the Euler-Lagrange equations corresponding to the variational principle (1.4). It is clear from their derivation that they do not depend on the choice of coordinates, and that they reduce to Newton's equations when L is the kinetic potential and the coordinates are Cartesian.

3.2 ENDPOINT VARIATIONS

One may consider a wider class of comparison functions for which the varied motion does not begin or end at the same point or time as the unvaried motion.

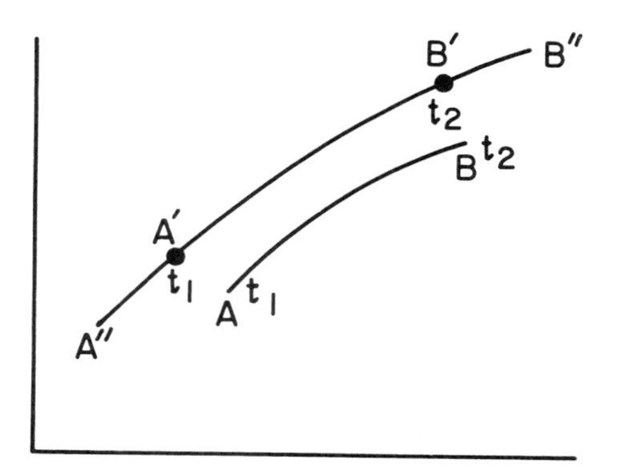

Fig. 2.1 Variations at Endpoints. A and A' are both passed at the time t_1. B and B' are both reached at the time t_2.

Then

$$q_k(B'') \neq q_k(B)$$

$$t(B'') \neq t(B) \qquad (2.1)$$

and

$$q_k(A'') \neq q_k(A)$$

$$t(A'') \neq t(A) \quad . \qquad (2.2)$$

Consider

$$\Delta q_k(B) \equiv q_k(B'') - q_k(B)$$

$$= [q_k(B') - q_k(B)] + [q_k(B'') - q_k(B')]$$

or

$$\Delta q_k(B) = \delta q_k(B) + \dot{q}_k(B)\, \Delta t_B \quad , \tag{2.3}$$

where

$$\Delta t_B \equiv t(B'') - t(B') = t(B'') - t(B) \quad . \tag{2.4}$$

In a variation of this kind, Eq. (1.10) also acquires the additional term L Δt, and therefore altogether we have

$$\delta S = \int_{t_1}^{t_2} \sum_k \left[\frac{\partial L}{\partial q_k} - \frac{d}{dt}\left(\frac{\partial L}{\partial \dot{q}_k}\right) \right] \delta q_k(t)\, dt$$

$$+ \left[L\, \Delta t + \sum_k \frac{\partial L}{\partial \dot{q}_k} \delta q_k \right]_{t_1}^{t_2} \quad . \tag{2.5}$$

If one now substitutes (2.3) in the integrated bracket of (2.5) he finds

$$\left[L\, \Delta t + \sum_k \frac{\partial L}{\partial \dot{q}_k} \delta q_k \right]_{t_1}^{t_2} = \left[\left[L - \sum \frac{\partial L}{\partial \dot{q}_k} \dot{q}_k \right] \Delta t + \right.$$

$$\left. + \sum_k \left(\frac{\partial L}{\partial \dot{q}_k}\right)_B \Delta q_k \right]_A^B \quad . \tag{2.6}$$

Then the formula for the most general variation is

$$\delta S = \int_{t_1}^{t_2} \sum_k \left[\frac{\partial L}{\partial q_k} - \frac{d}{dt}\left(\frac{\partial L}{\partial \dot{q}_k}\right) \right] \delta q_k \, dt + $$

$$+ \left[\sum_k p_k \, \Delta q_k - H \, \Delta t \right]_A^B \tag{2.7}$$

where

$$p_k = \frac{\partial L}{\partial \dot{q}_k} \tag{2.8}$$

$$H = \sum_k p_k \, \dot{q}_k - L \quad . \tag{2.9}$$

The conjugate momentum p_k and the Hamiltonian H thus appear naturally when the endpoint is varied.

Hamilton-Jacobi Equation

If the varied paths all satisfy the equations of motion (1.14) but differ in the positions and times of their end-points, then it follows from (2.7) that

$$\Delta S = \left[\sum_k p_k \, \Delta q_k - H \, \Delta t \right]_A^B \tag{2.10}$$

or

$$p_k = \frac{\partial S}{\partial q_k} \tag{2.11}$$

$$H = -\frac{\partial S}{\partial t} \quad . \tag{2.12}$$

Finally (2.11) and (2.12) may be combined to yield the Hamilton-Jacobi equation.

$$H\left(q_1 \cdots q_f \frac{\partial S}{\partial q_1} \cdots \frac{\partial S}{\partial q_f}\right) = -\frac{\partial S}{\partial t} \quad . \tag{2.13}$$

Equation (2.13) is a single partial differential equation for the function $S(q_1 \cdots q_f, t)$.

Invariance Group

The Lagrangian is invariant under the group of translations and rotations. It follows from (2.10) that, provided the equations of motion are satisfied

$$\left[\sum_k p_k \, \Delta q_k - H \, \Delta t\right]_A^B = 0 \tag{2.14}$$

when Δq_k and Δt are induced by translations and rotations. If we choose Cartesian coordinates $(x_{i\alpha})$ there are the following separate cases:

a) Translations: $\Delta t = 0, \quad \Delta x_{i\alpha} = \Delta x \qquad \alpha = 1,2,3$

$$\sum_i p_{i\alpha}^A \, \Delta x_{i\alpha} = \sum_i p_{i\alpha}^B \, \Delta x_{i\alpha}$$

$$\therefore \sum_i p_{i\alpha}^A = \sum_i p_{i\alpha}^B$$

or

$$\underset{\sim}{P}^A = \underset{\sim}{P}^B \quad . \tag{2.15}$$

b) Time displacement:

$$E^A = E^B \quad . \tag{2.16}$$

c) Rotation: $\Delta t = 0, \quad \Delta x_{i\alpha} = \sum \varepsilon_{\alpha\beta\gamma} x_{i\beta} \Delta\omega_\gamma$

or

$$\left[\sum \varepsilon_{\alpha\beta\gamma} x_{i\beta} p_{i\gamma} \right]_B = \left[\sum \varepsilon_{\alpha\beta\gamma} x_{i\beta} p_{i\gamma} \right]_A \tag{2.17}$$

$$\therefore L_\alpha^A = L_\alpha^B \quad .$$

In this way one obtains the conservation laws from the invariance group again.[2]

Notice that the Lagrangian, like the Hamiltonian, is not invariant under Galilean transformations.[3]

<u>Principle of Least Action (Principle of Maupertuis)</u>

Integrate (2.10) for a conservative system. Then

$$S = \int_{t_1}^{t_2} \sum_s p_s \dot{q}_s \, dt - E(t_2 - t_1)$$

$$= S_o - E(t_2 - t_1) \tag{2.18}$$

where

$$S_o = \sum_s \int_{t_1}^{t_2} p_s \, dq_s \quad . \tag{2.19}$$

Consider a varied motion that has the same energy and endpoints, but not the same initial and final times, as the actual motion. Then by (2.10) the change in the total action in such a variation is

$$\Delta S = - E \, \Delta t \tag{2.20}$$

and by (2.18)

$$\Delta S = \Delta S_o - E \, \Delta t \quad . \tag{2.21}$$

Therefore

$$\Delta S_o = 0 \quad . \tag{2.22}$$

It follows that S_o is stationary with respect to all motions with the same energy and endpoints when one varies the total transit time. In Hamilton's principle, on the other hand, one varies the energy while holding the time and endpoints fixed.

We may, following Landau, refer to S as the action, to Hamilton's principle as the principle of least action, to S_o

as the abbreviated action, and to (2.22) as the principle of Maupertuis. Other authors, particularly earlier ones, refer to S_o as the action and to (2.22) as the principle of least action.

Let the Lagrangian have the following form

$$L = \frac{1}{2} \sum a_{mn} \dot{q}_m \dot{q}_n - V(q\cdots) \quad . \tag{2.23}$$

Then the kinetic energy is

$$T = \frac{1}{2} \sum p_m \dot{q}_m \tag{2.24}$$

and therefore (2.22) implies

$$\Delta \int_{t_1}^{t_2} T \, dt = 0 \quad . \tag{2.25}$$

3.3 HAMILTON'S EQUATIONS

Hamilton's equations may be obtained from Hamilton's principle just as Lagrange's equations are. It is only necessary to express the variational principle in terms of the independent variables p_k and q_k and the Hamiltonian function as follows:

$$S = \int_{t_1}^{t_2} L\, dt = \int_{t_1}^{t_2} [\sum p_k \dot{q}_k - H(p,q)]dt \quad . \qquad (3.1)$$

One then varies the q_k and p_k independently:

$$\delta S = \int_{t_1}^{t_2} \sum_k (\delta p_k \dot{q}_k + p_k \delta\dot{q}_k - \frac{\partial H}{\partial p_k}\delta p_k - \frac{\partial H}{\partial q_k}\delta q_k)\, dt \quad , \qquad (3.2)$$

but

$$\int_{t_1}^{t_2} p_k \delta\dot{q}_k\, dt = \int_{t_1}^{t_2} p_k \frac{d}{dt}\delta q_k\, dt$$

$$= \Big[p_k \delta q_k\Big]_{t_1}^{t_2} - \int_{t_1}^{t_2} \delta q_k \dot{p}_k\, dt \quad . \qquad (3.3)$$

Under the conditions of Hamilton's principle, namely, $\delta q_k(t_2) = \delta q_k(t_1) = 0$, one has by combining (3.3) and (3.2)

$$\delta S = \int_{t_1}^{t_2} \sum_k \left[\delta p_k \left(\dot{q}_k - \frac{\partial H}{\partial p_k}\right) - \delta q_k \left(\dot{p}_k + \frac{\partial H}{\partial q_k}\right) \right] dt = 0 \; .$$

Since the variations δp_k and δq_k are independent and arbitrary, we may infer:

$$\dot{q}_k = \frac{\partial H}{\partial p_k}$$

$$\dot{p}_k = -\frac{\partial H}{\partial q_k} \quad . \tag{3.5}$$

These are Hamilton's equations.

Finally, notice that the variational principle, (1.3) and (1.4), may be generalized by adding to the integrand a perfect differential which is not varied at the endpoints

$$S' = \int_{t_1}^{t_2} \left(L + \frac{dF}{dt}\right) dt = \int_{t_1}^{t_2} L\, dt + F(2) - F(1)$$

$$S' = S + F(2) - F(1) \tag{3.6}$$

and

$$\delta S' = \delta S \tag{3.7}$$

since $\delta F(2) = \delta F(1) = 0$ by hypothesis.

Therefore, the variational principle leading to (3.5) may also be written

$$\delta \int_{t_1}^{t_2} \left[\sum p_k \dot{q}_k - H(p,q) + \frac{dF}{dt} \right] dt = 0 \tag{3.8}$$

where

$$\delta F(2) = \delta F(1) = 0 \quad . \tag{3.8a}$$

In other words the variational principle is invariant under the additive transformations of the integrand described by (3.8). We shall now see that the resulting arbitrariness in the integrand corresponds exactly to the possibility of making contact transformations on Hamilton's equations. Since the sum of two perfect differentials is again a perfect differential, these transformations, and therefore the contact transformations, have the closure property. In addition there is an identity, every additive transformation of this kind has an inverse, and the associative law holds. Consequently these transformations and therefore the contact transformations form a group.

3.4 CONTACT TRANSFORMATIONS

The transformations under which Hamilton's equations are invariant are called contact transformations and they may be regarded as point transformations in phase space as follows:

$$P_k = P_k(p, q, t)$$

$$Q_k = Q_k(p, q, t) \tag{4.1}$$

such that

$$\dot{Q}_k = \frac{\partial K}{\partial P_k}$$

$$\dot{P}_k = -\frac{\partial K}{\partial Q_k} \tag{4.2}$$

where K is the new Hamiltonian. Equations (4.2) may be obtained from the variational principle (3.1)

$$\delta \int_{t_1}^{t_2} [\sum P_k \dot{Q}_k - K(P,Q)] \, dt = 0 \quad . \tag{4.3}$$

According to the remark made at the end of the preceding section after Eq. (3.8), it follows that Eq. (4.3) describes the same dynamical system as (3.8) if the integrands differ only by a perfect differential which is not varied at the endpoints. That is, $H(p,q)$ describes the same system as $K(P,Q)$ if

$$\sum p_k \dot{q}_k - H = \sum P_k \dot{Q}_k - K + \frac{dF}{dt} \tag{4.4}$$

or

$$dF = \sum (p_k dq_k - P_k dQ_k) + (K - H) \, dt \tag{4.4a}$$

and

$$\frac{\partial F}{\partial q_k} = p_k \tag{4.5a}$$

$$\frac{\partial F}{\partial Q_k} = - P_k \tag{4.5b}$$

$$\frac{\partial F}{\partial t} = K - H \quad . \tag{4.5c}$$

Equations (4.5) then determine a contact transformation from (3.5) to (4.2) in terms of the function $F(q,Q,t)$. Then $F(q,Q,t)$ is called the generating function for the contact transformation.

The foregoing remarks remain true if dF is replaced by $d\tilde{F}$ defined as follows:

$$dF = d(\tilde{F} - \sum P_k Q_k) \quad . \tag{4.6}$$

Then

$$d\tilde{F} = \sum_k (p_k dq_k + Q_k dP_k) + (K - H)\, dt \tag{4.7}$$

and

$$p_k = \frac{\partial \tilde{F}}{\partial q_k} (q, P, t) \tag{4.8a}$$

$$Q_k = \frac{\partial \tilde{F}}{\partial P_k} (q, P, t) \tag{4.8b}$$

$$K - H = \frac{\partial \tilde{F}}{\partial t}(q, P, t) \quad . \tag{4.8c}$$

By appropriate Legendre transformations one may obtain the following four kinds of generating function:

$$F(q,Q,t) \;, \quad \tilde{F}(q,P,t) \;, \quad F'(p,Q,t) \;, \quad F''(p,P,t) \quad . \tag{4.9}$$

Examples:

(a) $$F = \sum q_s Q_s \tag{4.10}$$

$$p_k = \frac{\partial F}{\partial q_k} = Q_k \tag{4.11a}$$

$$P_k = -\frac{\partial F}{\partial Q_k} = -q_k \quad . \tag{4.11b}$$

This transformation is referred to in Chapter I [Eq. (2.13)].

(b) $$\tilde{F} = \sum q_s P_s \tag{4.12}$$

$$p_k = \frac{\partial \tilde{F}}{\partial q_k} = P_k \tag{4.12a}$$

$$Q_k = \frac{\partial \tilde{F}}{\partial P_k} = q_k \quad . \tag{4.12b}$$

This is the identity transformation.

$$\text{(c)} \quad \tilde{F} = \sum q_s P_s + \varepsilon G(q,P) \tag{4.13}$$

where ε is infinitesimal.

$$p_k = P_k + \varepsilon \frac{\partial G}{\partial q_k} (q,P)$$

$$Q_k = q_k + \varepsilon \frac{\partial G}{\partial P_k} (q,P) \quad .$$

Then to terms of order ε

$$P_k = p_k - \varepsilon \frac{\partial G}{\partial q_k} (q,p) = p_k - \varepsilon[G,p_k] \tag{4.14a}$$

$$Q_k = q_k + \varepsilon \frac{\partial G}{\partial p_k} (q,p) = q_k - \varepsilon\,[G,q_k] \quad . \tag{4.14b}$$

Equations (4.14) represent an infinitesimal contact transformation, and correspond to Eqs. (5.15) of Chapter 2, which describe an infinitesimal unitary transformation.

The correspondence is established as usual by the substitution

$$[\ ,\ ] \rightarrow \frac{(\ ,\ )}{i\hbar} \quad . \tag{4.15}$$

3.5 POISSON BRACKETS AND CONTACT TRANSFORMATIONS

In order to discuss contact transformations as point transformations in phase space, it is useful to introduce the following notation:

$$x^k = \begin{bmatrix} q_1 \\ p_1 \\ q_2 \\ p_2 \\ \vdots \end{bmatrix} \tag{5.1}$$

Hamilton's equations may then be written in the following form:

$$\begin{bmatrix} \dot{q}_1 \\ \dot{p}_1 \\ \dot{q}_2 \\ \dot{p}_2 \\ \vdots \end{bmatrix} = \begin{bmatrix} 0 & 1 & 0 & 0 & \cdots \\ -1 & 0 & 0 & 0 & \\ 0 & 0 & 0 & 1 & \\ 0 & 0 & -1 & 0 & \\ & \vdots & & & \ddots \end{bmatrix} \begin{bmatrix} \frac{\partial H}{\partial q_1} \\ \frac{\partial H}{\partial p_1} \\ \frac{\partial H}{\partial q_2} \\ \frac{\partial H}{\partial p_2} \\ \vdots \end{bmatrix} \tag{5.2}$$

or simply

$$\dot{x}^k = \varepsilon^{k\ell} \frac{\partial H}{\partial x^\ell} \tag{5.3}$$

where

$$\varepsilon^{k\ell} = \begin{bmatrix} 0 & 1 & & & \\ -1 & 0 & & & \\ & & 0 & 1 & \\ & & -1 & 0 & \\ & & & & \ddots \end{bmatrix} \tag{5.4}$$

In this notation an arbitrary transformation from (q_k, p_k) to (q_k', p_k') implies

$$dx'^k = \frac{\partial x'^k}{\partial x^m} dx^m \tag{5.5}$$

and

$$\frac{dx'^k}{dt} = \frac{\partial x'^k}{\partial x^m} \frac{dx^m}{dt} \quad . \tag{5.6}$$

Combining (5.6) and (5.3) one has

$$\begin{aligned} \dot{x}'^k &= \frac{\partial x'^k}{\partial x^m} \dot{x}^m = \frac{\partial x'^k}{\partial x^m} \varepsilon^{mn} \frac{\partial H}{\partial x^n} \\ &= \left(\frac{\partial x'^k}{\partial x^m} \varepsilon^{mn} \frac{\partial x'^\ell}{\partial x^n} \right) \frac{\partial H}{\partial x'^\ell} \end{aligned} \tag{5.7}$$

where we have restricted ourselves to the case $H'(x') = H(x)$.

The condition that (5.3) and (5.7) have the same form is then

$$\varepsilon^{k\ell} = \frac{\partial x'^k}{\partial x^m} \varepsilon^{mn} \frac{\partial x'^\ell}{\partial x^n} \quad . \tag{5.8}$$

The preceding equations (5.8) are therefore the conditions for a contact transformation. They may be simply expressed in terms of Poisson brackets, for

$$\begin{aligned} \frac{\partial A}{\partial x^m} \varepsilon^{mn} \frac{\partial B}{\partial x^n} &= \frac{\partial A}{\partial x^1} \varepsilon^{12} \frac{\partial B}{\partial x^2} + \frac{\partial A}{\partial x^2} \varepsilon^{21} \frac{\partial B}{\partial x^1} + \cdots \\ &= \frac{\partial A}{\partial q_1} \frac{\partial B}{\partial p_1} - \frac{\partial A}{\partial p_1} \frac{\partial B}{\partial q_1} + \cdots \\ &= [A, B] \quad . \end{aligned} \tag{5.9}$$

Therefore (5.8) becomes

$$[x'^k, x'^\ell] = \varepsilon^{k\ell} \tag{5.10}$$

or

$$\begin{aligned} [x'^{2s-1}, x'^{2t}] &= [q_s', p_t'] \\ &= \sum_m \left(\frac{\partial q_s'}{\partial q_m} \frac{\partial p_t'}{\partial p_m} - \frac{\partial q_s'}{\partial p_m} \frac{\partial p_t'}{\partial q_m} \right) = \delta_{st} \end{aligned}$$

and all other kinds of bracket vanish.

These are the conditions quoted in Chapter I [Eqs. (3.16)].[4] Moreover (5.10) are only a special case of the statement that [A, B] is invariant under a contact transformation. For

$$[A, B]' = \frac{\partial A}{\partial x'^m} \varepsilon^{mn} \frac{\partial B}{\partial x'^n}$$

$$= \frac{\partial x^s}{\partial x'^m} \frac{\partial A}{\partial x^s} \varepsilon^{mn} \frac{\partial B}{\partial x^t} \frac{\partial x^t}{\partial x'^n}$$

$$= \left(\frac{\partial A}{\partial x^s}\right) \left(\frac{\partial x^s}{\partial x'^m} \varepsilon^{mn} \frac{\partial x^t}{\partial x'^n}\right) \left(\frac{\partial B}{\partial x^t}\right)$$

$$= \left(\frac{\partial A}{\partial x^s}\right) \varepsilon^{st} \left(\frac{\partial B}{\partial x^t}\right)$$

by (5.8).

Therefore,

$$[A, B]' = [A, B] \quad . \tag{5.11}$$

Transformations satisfying (5.8), i.e., canonical transformations, are also known as symplectic. The Poisson bracket is then a symplectic invariant.[5]

3.6 THE HAMILTON-JACOBI EQUATION

According to Eq. (4.5c) one may choose the generating function F(q,Q,t) of a contact transformation so that

$$\frac{\partial F}{\partial t}(q,Q,t) + H(p,q) = K = 0 \quad . \tag{6.1}$$

If it is possible to find the F(q,Q,t) which satisfies this condition, then by (4.2)

$$\dot{P}_k = \dot{Q}_k = 0 \quad . \tag{6.2}$$

The P_k and Q_k arrived at in this way are constants, and the Eqs. (4.5) provide a solution of the dynamical problem by determining p_k and q_k at any time t in terms of the 2f constants of integration P_k and Q_k. That is, Eq. (4.5b) connects q_k with Q_k, P_k and t is therefore the equation of the orbit in terms of the 2f boundary conditions:

$$q_k = q_k(Q, P, t) \quad . \tag{6.3a}$$

Likewise Eq. (4.5a) connects p_k with q_k, Q_k, and t, and therefore determines the momentum at any point in the orbit:

$$p_k = p_k(q, Q, t) \quad . \tag{6.3b}$$

To obtain the solution in this form one must determine F(q,Q,t) from (6.1). This condition may be reformulated as a partial differential equation by using (4.5a). Then

$$\frac{\partial F}{\partial t}(q,Q,t) + H\left(\frac{\partial F}{\partial q}(q,Q,t),\ q,\ t\right) = 0 \quad . \tag{6.4a}$$

This is a non-linear partial differential equation involving all the partial derivatives $\partial F/\partial q_k$ and $\partial F/\partial t$ in general. Equation (6.4) is in fact the same as (2.13) of this chapter and is therefore solved by

$$S = \int_{Q_1 Q_2 \cdots}^{q_1 q_2 \cdots} L\, dt = F(q,Q,t) \quad . \tag{6.5}$$

One can therefore formally solve the Hamilton-Jacobi equation by carrying out this integration along the actual motion.

The actual dynamical motion induces a contact transformation between (Q,P) at the initial time and (q,p) at a different time. As the two endpoints approach, the contact transformation becomes infinitesimal and the connection between (Q,P) and (q,p) is given simply by Hamilton's equations. Hamilton's equations therefore describe an infinitesimal contact transformation; or, as it is sometimes stated, the history of a dynamical system may be regarded as the gradual unfolding of a contact transformation.

The preceding discussion and equation (6.4a) is based on the function F(q,Q,t), but the Hamilton-Jacobi equation also holds for $\tilde{F}(q,P,t)$ defined by (4.6). That is, by (4.8)

$$H\left(q, \frac{\partial \tilde{F}}{\partial q}(q,P,t)\right) + \frac{\partial \tilde{F}}{\partial t}(q,P,t) = 0 \quad . \tag{6.4b}$$

(For the other forms of the generating function and the Hamilton-Jacobi equation, see problem 13 at the end of this chapter.)

(a) Isolated Systems

The kind of physical system considered in this book is further restricted by the condition that

$$\frac{\partial H}{\partial t} = 0 \quad .$$

In this case one can put

$$\tilde{F}(q,P,t) = W(q,P) - \alpha_1 t \quad . \tag{6.6}$$

Then the Hamilton-Jacobi equation (6.4b) becomes

$$H\left(\frac{\partial W}{\partial q}, q\right) = \alpha_1 \quad . \tag{6.7}$$

According to the Hamilton-Jacobi prescription, one solves (6.7) for any complete integral $W(q_1 \cdots q_f \; \alpha_1 \cdots \alpha_{f+1})$ containing f constants of integration in addition to the energy which is denoted by α_1. Information about the orbit is then obtained by differentiating $W(q_1 \cdots q_f \; \alpha_1 \cdots \alpha_{f+1})$ with respect to the constants of integration (α_k) in a way determined by (4.8), namely:

$$p_k = \frac{\partial W}{\partial q_k} \tag{6.8a}$$

$$\beta_1 = \frac{\partial W}{\partial \alpha_1} - t \tag{6.8b}$$

$$\beta_k = \frac{\partial W}{\partial \alpha_k} \qquad k = 2 \cdots f+1 \tag{6.8c}$$

where the substitution $\tilde{F} = W - \alpha_1 t$ has been made in (4.8). Here the constant momenta (P_k) have been chosen to be the energy, α_1, and the initial momenta, α_k $(k \neq 1)$; the coordinates (Q_k) conjugate to the α_k have been denoted by β_k.

Then the orbit is described by (6.8c) while the connection between position in the orbit and the time is given by (6.8b). The momenta at any point in the orbit are given by (6.8a). The constants (α_k) may be chosen to be integrals of the motion like the angular momentum.

(b) Contact Transformation Defined by $W(q_1 \cdots \alpha_1 \cdots)$

We also note that the time independent function $W(q_1 \cdots \alpha_1 \cdots)$ defines a contact transformation that is quite different from the equations (6.8) which are based on (4.8) and the time dependent generator $\tilde{F}(q_1 \cdots \alpha_1 \cdots t)$. This different transformation may be obtained by going back to (4.8) and specializing $\tilde{F}$ to be a function $W(q,P)$ that does not depend on the time, so that

$$p_k = \frac{\partial W}{\partial q_k}(q,P) \tag{6.9a}$$

$$Q_k = \frac{\partial W}{\partial P_k}(q,P) \tag{6.9b}$$

$$H = K(P_1 P_2 \cdots) \tag{6.9c}$$

where K is chosen to be independent of all the Q_k. As in the preceding case (a) the new momenta (P_k) are chosen to be constants of the motion. In contrast to the preceding case, however, the new Hamiltonian does not vanish and in addition $K \neq P_1$. Therefore the Q_k are not constants as they were in (6.8). Instead

$$\dot{Q}_k = \frac{\partial K}{\partial P_k}(P_1 \cdots) = \nu_k = \text{constant} \tag{6.10}$$

and

$$Q_k = \nu_k t + \gamma_k \quad . \tag{6.10a}$$

This kind of transformation is used in the method of action and angle variables, which will be described in Chapter 5 in connection with our discussion of orbits.

(c) Wave Front Associated With a Classical System

By means of an equation like (6.6) it is also possible to associate a moving surface, or wavefront, with the evolution of the dynamical system. Consider surfaces of constant

$S(q,Q,t)$

$$S(q,Q,t) = W(q,Q) - Et \quad .$$

Here the variables are (q,Q) in contrast to (6.6) where they are (q,P). The surfaces $W(q,Q)$ are fixed hypersurfaces in q-space for given Q. A constant value of S will therefore propagate on a moving hypersurface as follows:

$$S(q_1,Q_1t_1) = S(q_2,Q_1t_2)$$

$$W(q_1,Q) - Et_1 = W(q_2,Q) - Et_2$$

$$W(q_2,Q) = W(q_1,Q) + E(t_2-t_1) \quad . \tag{6.11}$$

This equation determines the time, t_2-t_1, required for the wavefront to move from its position at t_1 to its position at t_2. One may recover the usual information about the trajectories from knowledge of this moving surface, for the trajectories are orthogonal to the wavefront and satisfy

$$p_k = \frac{\partial S}{\partial q_k} \tag{6.11a}$$

$$E = -\frac{\partial S}{\partial t} \quad . \tag{6.11b}$$

The wavefront changes shape as it moves, since the velocity varies from point to point in a way determined by (6.11). That is

$$\sum \frac{\partial W}{\partial q^k} \Delta q^k = E \, \Delta t$$

or

$$|\nabla W| \, \Delta q = E \, \Delta t$$

where $|\nabla W|$ is the magnitude of the gradient of W in the many dimensional q-space, and where Δq is measured normal to the W-hypersurface. Then the wave velocity is

$$u = \frac{dq}{dt} = \frac{E}{|\nabla W|} \quad . \tag{6.12a}$$

But

$$|\nabla_k W| = |\nabla_k S| = |p_k| \equiv p$$

and therefore

$$u = E/p \quad . \tag{6.12b}$$

The velocity of the wavefront at a point is then inversely related to the velocity associated with the trajectory at the same point. If the dynamical system consists of a single particle, then the wave moves in three dimensional space in such a way that the wavefront is always perpendicular to the orbit of the particle.

In this way Hamilton was led, as long ago as 1834, to

associate a wave motion with mechanical systems in order to carry over the concepts of waves and rays which were known to hold for light. Although one can thus, without going beyond classical theory, associate a wavefront with the dynamical system, it is not possible to introduce a wave length also, since the latter requires Planck's constant. If one does add Planck's relation, $E = h\nu$, then (6.12b) becomes the de Broglie relation

$$\lambda = \frac{u}{\nu} = \frac{h}{p} \quad . \tag{6.13}$$

We now know that classical mechanics is an approximation to wave mechanics that is valid in the limit of short waves just as geometrical optics is the corresponding limit for light. The further development of this idea leads to the eikonal approximation.

3.7 THE SCHRÖDINGER EQUATION AND THE HAMILTON-JACOBI EQUATION

In the preceding section the Hamilton-Jacobi equation was obtained by making a contact transformation to canonical coordinates which are independent of the time. In (2.6) of Chapter 2, the Schrödinger equation was similarly obtained by making a unitary transformation to a basis in which the physical operators are time independent. We may summarize in the following table.

TABLE 7.1

Classical	Quantum
Hamilton	**Heisenberg**
$\dot{q}_k = [q_k, H]$	$i\hbar\, \dot{q}_k = (q_k, H)$
$\dot{p}_k = [p_k, H]$	$i\hbar\, \dot{p}_k = (p_k, H)$
	$\frac{d}{dt} \vert\rangle = 0$
Contact Transformation	**Unitary Transformation**
$p_k = \frac{\partial \tilde{F}}{\partial q_k}$	$P_k = U\, p_k\, U^{-1}$
$Q_k = \frac{\partial \tilde{F}}{\partial P_k}$	$Q_k = U\, q_k\, U^{-1}$
$K = H + \frac{\partial \tilde{F}}{\partial t}$	$K = U\, H\, U^{-1} + i\hbar\, U \frac{d}{dt} U^{-1}$
Transformed Equations	**Transformed Equations**
$\dot{Q}_k = [Q_k, K]$	$i\hbar\, \dot{Q}_k = (Q_k, K)$
$\dot{P}_k = [P_k, K]$	$i\hbar\, \dot{P}_k = (P_k, K)$
Hamilton-Jacobi (K = 0)	**Schrödinger (K = 0)**
$\dot{Q}_k = 0$	$\dot{Q}_k = 0$
$\dot{P}_k = 0$	$\dot{P}_k = 0$
$H\left(\frac{\partial \tilde{F}}{\partial q_1} \cdots q_1 \cdots\right) + \frac{\partial \tilde{F}}{\partial t} = 0$	$H\left(\frac{\hbar}{i} \frac{\partial}{\partial q_1} \cdots q_1 \cdots\right) \vert\rangle + \frac{\hbar}{i} \frac{\partial \vert\rangle}{\partial t} = 0$

As was claimed in (1.19) of Chapter 2, Hamilton's classical equations emerge as the correspondence limit of the Heisenberg equations since

$$(A,B) = [A,B](i\hbar) + O(\hbar^2) \quad .$$

The correspondence between unitary quantum transformations and classical contact transformations has been pointed out earlier by comparing (4.14) of this chapter with (5.15) of Chapter 2. There we put

$$\tilde{F} = \sum q_k P_k + \varepsilon G \quad , \text{ where } \quad G = \left(\frac{\partial \tilde{F}}{\partial \varepsilon}\right)_0 \quad , \tag{7.1}$$

and found classical equations (4.14) that may be rewritten as follows:

$$\delta p_k = [p_k, \varepsilon G] \tag{7.2a}$$

$$\delta q_k = [q_k, \varepsilon G] \tag{7.2b}$$

$$\delta H = \frac{\partial}{\partial t} (\varepsilon G) \quad . \tag{7.2c}$$

Let the unitary transformation appearing in the table 7.1 be

$$U = e^{(i/\hbar)\varepsilon G} \quad , \quad \text{where} \quad G = \frac{\hbar}{i}\left(\frac{\partial U}{\partial \varepsilon}\right)_0 \quad .$$

Then the quantum formulas appearing in the table may be put in the following infinitesimal form:

$$\delta p_k = \frac{1}{i\hbar} (p_k, \varepsilon G) \tag{7.2a)'}$$

$$\delta q_k = \frac{1}{i\hbar} (q_k, \varepsilon G) \tag{7.2b)'}$$

$$\delta H = \frac{1}{i\hbar} (H, \varepsilon G) + \frac{d}{dt} (\varepsilon G) = \frac{\partial}{\partial t} (\varepsilon G) \tag{7.2c)'}$$

where (1.14) of Chapter 2 has been used to reduce δH. Then (7.2) and (7.2)' correspond precisely. To obtain either the Hamilton-Jacobi or the Schrödinger equation the contact or unitary transformation must be continued until the new Hamiltonian vanishes.

As already discussed, the solution of the Hamilton-Jacobi equation provides a solution of the dynamical problem in terms of equations like

$$Q_k = \frac{\partial \tilde{F}}{\partial P_k} (q, P, t)$$

that describe the contact transformation in table 7.1. In an exactly similar way the solution of the Schrödinger equation determines the unitary transformation, U, in this table and U provides a solution to the quantum problem according to

$$Q_k = U\, q_k\, U^{-1} \quad , \quad \text{or} \quad q_k = U^{-1}\, Q_k\, U \quad .$$

According to the preceding table the Hamilton-Jacobi

equation ought to be the classical limit of the Schrödinger equation. This point may be checked by the following method. Let

$$\psi = R\, e^{(i/\hbar)S} \tag{7.3}$$

where R is real and $S/\hbar$ is the phase of ψ. Then

$$\frac{\hbar}{i}\frac{\partial}{\partial q}\, e^{(i/\hbar)S} \cdot R = e^{(i/\hbar)S} \left[\left(\frac{\hbar}{i}\frac{\partial}{\partial q} + \frac{\partial S}{\partial q} \right) R \right] . \tag{7.4}$$

The right side of (7.4) is again of the form $e^{(i/\hbar)S}R$ where the new R is the square bracket of (7.4). Therefore

$$\left(\frac{\hbar}{i}\frac{\partial}{\partial q} \right)^2 e^{(i/\hbar)S} \cdot R = e^{(i/\hbar)S} \left[\left(\frac{\hbar}{i}\frac{\partial}{\partial q} + \frac{\partial S}{\partial q} \right)^2 R \right] . \tag{7.5}$$

Similarly

$$\left(-\frac{\hbar}{i}\frac{\partial}{\partial t} \right) e^{(i/\hbar)S}\, R = e^{(i/\hbar)S} \left[\left(-\frac{\hbar}{i}\frac{\partial}{\partial t} - \frac{\partial S}{\partial t} \right) R \right] . \tag{7.6}$$

It follows that

$$H\left(\frac{\hbar}{i}\frac{\partial}{\partial q_1} \ldots ; q_1 \ldots \right) \psi = -\frac{\hbar}{i}\frac{\partial \psi}{\partial t} \tag{7.7}$$

becomes

$$e^{(i/\hbar)S}\, H\left[\frac{\hbar}{i}\frac{\partial}{\partial q_1} + \frac{\partial S}{\partial q_1} \ldots ;\; q_1 \ldots\right] R =$$

$$= e^{(i/\hbar)S}\left(-\frac{\hbar}{i}\frac{\partial}{\partial t} - \frac{\partial S}{\partial t}\right) R \quad . \tag{7.8}$$

In the classical limit

$$\frac{\hbar}{i}\frac{\partial}{\partial q} << \frac{\partial S}{\partial q} \tag{7.9a}$$

$$\frac{\hbar}{i}\frac{\partial}{\partial t} << \frac{\partial S}{\partial t} \quad . \tag{7.9b}$$

If Eq. (7.8) is expanded in powers of $\hbar/S$, one obtains, up to terms of the first order, the following:[6]

$$\text{Terms of order } \left(\frac{\hbar}{S}\right)^0 : \quad H\left(\frac{\partial S}{\partial q_1} \cdots q_1 \cdots\right) = -\frac{\partial S}{\partial t} \tag{7.10}$$

$$\text{Terms of order } \left(\frac{\hbar}{S}\right)^1 : \quad \frac{\partial}{\partial t} R^2 + \sum \frac{\partial}{\partial q_k}\left(R^2 \frac{\partial H}{\partial p_k}\right) = 0 \quad . \tag{7.11}$$

(7.10) is the Hamilton-Jacobi equation and (7.11) may be read as the equation of continuity

$$\text{div}\,(\rho \underset{\sim}{v}) + \frac{\partial \rho}{\partial t} = 0 \tag{7.12}$$

where

$$\rho = R^2 = |\psi|^2 \tag{7.12a}$$

and

$$v_k = \frac{dq_k}{dt} = \frac{\partial H}{\partial p_k} \quad . \tag{7.12b}$$

Therefore to terms of the first order in $\frac{\hbar}{S}$ one obtains only a classical statistical theory. [Both (7.10) and (7.11) or (7.12) may be written down classically and of course neither equation contains $\hbar$.]

Example

The Schrödinger equation of a single particle in a potential is

$$-\frac{\hbar^2}{2m}\nabla^2\psi + V\psi = -\frac{\hbar}{i}\frac{\partial\psi}{\partial t} \quad . \tag{7.13}$$

Put $\psi = R\, e^{(i/\hbar)S}$. Then

$$-\frac{\hbar^2}{2m}\nabla^2\psi = e^{(i/\hbar)S}\frac{1}{2m}\left[\frac{\hbar}{i}\underset{\sim}{\nabla} + \underset{\sim}{\nabla}S\right]^2 R$$

and the Schrödinger equation now reads

$$\frac{1}{2m}\left[\left(\frac{\hbar}{i}\right)^2\nabla^2 + \frac{\hbar}{i}(\underset{\sim}{\nabla}S)\underset{\sim}{\nabla} + \left(\frac{\hbar}{i}\underset{\sim}{\nabla}\right)(\underset{\sim}{\nabla}S) + (\underset{\sim}{\nabla}S)^2\right]R + VR =$$

$$= \left[-\frac{\hbar}{i}\frac{\partial}{\partial t} - \frac{\partial S}{\partial t}\right]R \quad .$$

The terms linear in $\hbar$ are

$$\frac{1}{2m}\left[\frac{\hbar}{i}(\underset{\sim}{\nabla}S)(\underset{\sim}{\nabla}R) + \frac{\hbar}{i}\underset{\sim}{\nabla}(\underset{\sim}{\nabla}S\, R)\right] = -\frac{\hbar}{i}\frac{\partial R}{\partial t}$$

$$\frac{1}{2m}\left[2(\nabla S)(\nabla R) + R\nabla\ (\nabla S)\right] = -\frac{\partial R}{\partial t}$$

or

$$\frac{\partial R^2}{\partial t} + \nabla [R^2 \frac{1}{m} \nabla S] = 0 \quad . \tag{7.14a}$$

Again introduce $\rho = R^2$ and $v = \frac{1}{m}\nabla S$. Then

$$\frac{\partial \rho}{\partial t} + \text{div}\ (\rho v) = 0 \quad . \tag{7.14b}$$

If this equation of continuity is subtracted from (7.13) one gets

$$\left[\frac{1}{2m}(\nabla S)^2 + V - \frac{\hbar^2}{2m}\frac{\nabla^2 R}{R}\right] = -\frac{\partial S}{\partial t} \quad . \tag{7.15}$$

Equation (7.15) is the Hamilton-Jacobi equation with a modified potential

$$V - \frac{\hbar^2}{2m}\frac{\nabla^2 R}{R} \quad . \tag{7.16}$$

The additional potential

$$-\frac{\hbar^2}{2m}\frac{\nabla^2 R}{R} \tag{7.17}$$

is of order $\hbar^2$.

Efforts have been made to interpret quantum theory in terms of the classical theory and a quantum potential like

(7.17). Insofar as they agree with experiment, these attempts have merely paraphrased the more usual expositions of quantum theory.[7]

3.8 THE FORMAL SOLUTION OF SCHRÖDINGER'S EQUATION

In paragraph (3.1) it was shown how the equations of motion could be obtained from Hamilton's principle

$$\delta \int L\,dt = 0$$

and in (3.2) it was shown how the Hamilton-Jacobi equation appears when the endpoint is varied. Let us now rearrange and expand Table (7.1) to include these features, as well as their quantum analogues, although the latter have not yet been discussed.

The deepest formulation of classical dynamics is at the level of the action integral. The corresponding quantal formulation is based on the transformation function $\langle q't'|qt\rangle$ or transition amplitude which has therefore been put into correspondence with the action integral in the following table.

Let us next discuss the concept of the transition amplitude by formulating Schrödinger's equation in integral form.

By the general expansion theorem we have

$$|\,\rangle = \int |a'\rangle\langle a'|\,\rangle da'$$

TABLE 8.1

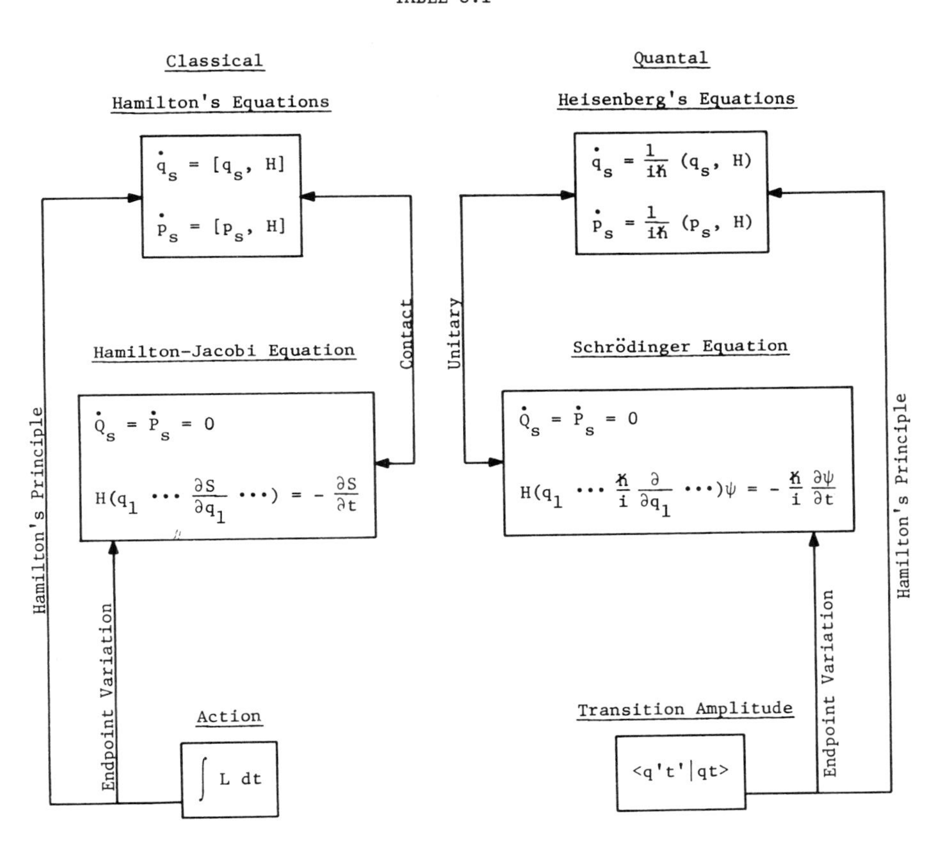

Classical
Hamilton's Equations
$\dot{q}_s = [q_s, H]$
$\dot{p}_s = [p_s, H]$
Hamilton-Jacobi Equation
$\dot{Q}_s = \dot{P}_s = 0$
$H(q_1 \cdots \frac{\partial S}{\partial q_1} \cdots) = -\frac{\partial S}{\partial t}$
Action
$\int L\, dt$
Hamilton's Principle
Endpoint Variation
Contact
Quantal
Heisenberg's Equations
$\dot{q}_s = \frac{1}{i\hbar}(q_s, H)$
$\dot{p}_s = \frac{1}{i\hbar}(p_s, H)$
Schrödinger Equation
$\dot{Q}_s = \dot{P}_s = 0$
$H(q_1 \cdots \frac{\hbar}{i}\frac{\partial}{\partial q_1} \cdots)\psi = -\frac{\hbar}{i}\frac{\partial \psi}{\partial t}$
Transition Amplitude
$\langle q't'|qt\rangle$
Unitary
Endpoint Variation
Hamilton's Principle

and in particular

$$| \ \rangle = \int |q't'\rangle \, dq' \, \langle q't'| \ \rangle$$

where q' stands for the complete set of generalized coordinates $(q_1' \ q_2' \ \cdots)$. Then

$$\langle q''t''| \ \rangle = \int \langle q''t''|q't'\rangle \, dq' \, \langle q't'| \ \rangle \tag{8.1}$$

or

$$\psi(q_1''\cdots t'') = \int K(q_1''\cdots t''; \ q_1'\cdots t') \times \psi(q_1'\cdots t') \, dq_1'\cdots \tag{8.2}$$

where

$$\psi(q't') = \langle q't'| \ \rangle \tag{8.3}$$

$$K(q''t''; \ q't') = \langle q''t''|q't'\rangle \quad . \tag{8.4}$$

If the kernel $K(q''t''; \ q't')$ is known, Eq. (8.2) must be equivalent to Schrödinger's equation since one can calculate $\psi(q''t'')$ from it if one knows $\psi(q't')$ at an earlier time. Then $K(q''t''; \ q't')$ (instead of the Hamiltonian) characterizes the dynamical problem in this formulation, and is known as the propagator.

If one has a complete set of solutions of the Schrödinger equation one may calculate this kernel as follows. Let

$$H\, U_n = E_n\, U_n \quad .$$

Then

$$\psi_n(q\ t) = \exp(-\frac{i}{\hbar} E_n t)\ U_n(q) \quad . \tag{8.5}$$

If (8.5) solves the Schrödinger equation then it must also solve (8.2) as follows

$$U_n(q'') \exp(-\frac{i}{\hbar} E_n t'') =$$
$$= \int K(q''t'';\ q't')\ U_n(q')\ \exp(-\frac{i}{\hbar} E_n t')\ dq'$$

or

$$\lambda_n U_n(q'') = \int K(q''t'';\ q't')\ U_n(q')\ dq' \tag{8.6}$$

where

$$\lambda_n = \exp\left(-\frac{i}{\hbar} E_n (t''-t')\right) \quad . \tag{8.6a}$$

The integral transformation described by (8.2) is of course a unitary transformation since it must preserve the probability integral; and its eigenvalues λ_n must therefore satisfy $|\lambda_n| = 1$.

If the set $U_n(q)$ is complete, then one may expand the kernel itself

$$K(q''t'';\ q't') = \sum C_m(q''t'';\ t')\ U_m^*(q') \quad .$$

Then

$$\lambda_n U_n(q'') = \sum C_m \int U^*_m(q')\ U_n(q')\ dq' = C_n$$

if the set $U_n(q)$ is also orthonormal, as we shall assume.

Then

$$K(q''t'';\ q't') = \sum \lambda_m U_m(q'')\ U^*_m(q') \tag{8.7}$$

$$= \sum \exp\left(-\frac{i}{\hbar} E_m(t''-t')\right) U_m(q'')\ U^*_m(q') \quad . \tag{8.7a}$$

Equation (8.7a) is then an explicit formula for the kernel in terms of the eigenfunctions and eigenvalues of the time independent Schrödinger equation.

In the Dirac notation, formula (8.7a) is obtained directly from (8.4) by inserting a complete set of intermediate states as follows

$$\langle q''t''|q't'\rangle = \sum_n \langle q''t''|n\rangle \langle n|q't'\rangle$$

$$= \sum_n \psi_n(q''t'')\ \psi_n(q't')^* \quad . \tag{8.8}$$

By (8.5) the preceding formula (8.8) is just the same as (8.7a).

The direct physical interpretation of

$$\langle q''t''|q't'\rangle$$

is obtained by substituting

$$\psi(q't') = \delta(q' - q_0') \quad . \tag{8.9}$$

Then

$$\psi(q''t'') = \langle q''t''|q_0't'\rangle \quad , \tag{8.10}$$

i.e., if the system at time t' is known to be precisely at point q_0' of configuration space, then at a later time t'' it has a probability amplitude $\langle q''t''|q_0't'\rangle$ of being at q''. It follows from (8.9) and (8.10) that

$$\lim_{t''\to t'} \langle q''t''|q't'\rangle = \delta(q''-q') \quad . \tag{8.11}$$

In addition it is clear from (8.8) that the transformation function must also satisfy the Schrödinger equation:

$$i\hbar \frac{\partial}{\partial t''} \langle q''t''|q't'\rangle = H\left(q'', \frac{\hbar}{i}\frac{\partial}{\partial q''}\right) \langle q''t''|q't'\rangle \tag{8.12}$$

$$= \langle q''t''|H|q't'\rangle \quad . \tag{8.12a}$$

Example (The Harmonic Oscillator)

The eigenfunctions of the oscillator are hermite functions and the energy levels are equally spaced. To evaluate (8.7) one may use the following formula

$$\sum_{n=0}^{\infty} U_n(x)\, U_n(y)\, \lambda^n =$$

$$= \frac{1}{\sqrt{\pi}} \frac{1}{[1-\lambda^2]^{1/2}} \exp\left[-\frac{1}{2}(x^2+y^2)\frac{1+\lambda^2}{1-\lambda^2} + \frac{2xy\lambda}{1-\lambda^2}\right] \quad (8.13)$$

where the U_n are the hermite functions.[8] Here λ_n of (8.6a) is

$$\lambda_n = \exp\left[-i\left(n+\frac{1}{2}\right)\omega(t''-t')\right] = \exp\left[-\frac{1}{2}i\omega(t''-t')\right]\lambda^n$$

where

$$\lambda = e^{-i\omega(t''-t')} \quad .$$

The evaluation of (8.13) is left as exercise 5 at the end of this chapter.

In solving the Schrödinger equation we have been led to the transition amplitude $\langle q'_2 t_2 | q'_1 t_1 \rangle$ which is a special case of $\langle b' t_2 | a' t_1 \rangle$ where a' and b' are arbitrary labelings of the initial and final states. In general $\langle b' t_2 | a' t_1 \rangle$ tells us the probability amplitude that a system prepared in state a', will be found subsequently in state b'. Therefore $\langle b' t_2 | a' t_1 \rangle$ answers the most general observational question that may be asked, and may be regarded as the basic formal object of the theory. If t_1 and t_2 are times sufficiently "before" and "after", then $\langle b'' t_2 | a' t_1 \rangle$ is an element of the S-matrix

$$\langle b''t_2|a't_1\rangle = \langle a''t_1|S|a't_1\rangle = \langle b''t_2|S|b't_2\rangle \quad .$$

In view of the preceding remarks it is natural to reconstruct the theory in such a way that the transition amplitude occupies the most fundamental position. This reconstruction has been carried out by Feynman and by Schwinger in two separate and complete formulations of quantum theory. Each of these versions can be developed *ab initio* from its own postulates and leads to the usual Heisenberg and Schrödinger formulations. In view of their fundamental nature it would be logically most satisfactory if both the Feynman and Schwinger formulations were developed without reference to the earlier versions of the quantum theory. That, however, is not what we would like to do here. Instead we are primarily interested in relating the different forms of quantum theory to each other and to classical theory in a rather loose but clear physical framework.

3.9 THE FEYNMAN SUM OVER PATHS[9]

According to Hamilton's principle in classical mechanics one calculates the action for all paths which satisfy the given boundary conditions in order to select that unique path for which

$$\delta S = 0 \quad .$$

That is, the class of varied functions is completely discarded at the end of this procedure.

In quantum mechanics, on the other hand, the actual motion is determined by the varied paths as well as by the unvaried path according to the Feynman principle:

$$\langle q't'|q\ t\rangle = \frac{1}{N} \sum_{\text{paths}} \exp\left[\frac{i}{\hbar} S(q't';\ qt)\right] \tag{9.1}$$

where the sum over paths is defined to be the sum over all conceivable motions including those which are classically discarded. We shall not discuss the general question of how the different paths are to be weighted; it will be sufficient to assign the same modulus (1/N) to all paths. As remarked in the preceding section, $\langle q't'|qt\rangle$ is the probability amplitude that the system be found at point q' and time t' if it is initially located at point q and time t.

The varied paths lying close to the classical motion interfere constructively since the phase S has an extremal at the classical motion. Therefore if $\hbar/S$ is small, one may approximate the transformation function $\langle q't'|qt\rangle$ by taking into account only those varied paths which lie in a narrow tube about the classical motion. As $\hbar/S \to 0$, the tube becomes

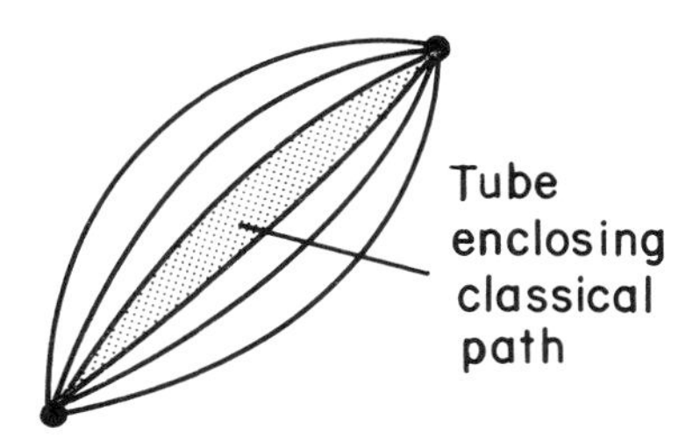

Figure 9.1 Paths contributing to $\langle q't'|qt\rangle$

narrower and approaches just the classical path in the correspondence limit. The elementary experiments that demonstrate wave particle duality also suggest a simple intuitive picture of the Feynman sum. Consider the diffraction pattern associated with a screen containing several openings (Fig. 9.2).

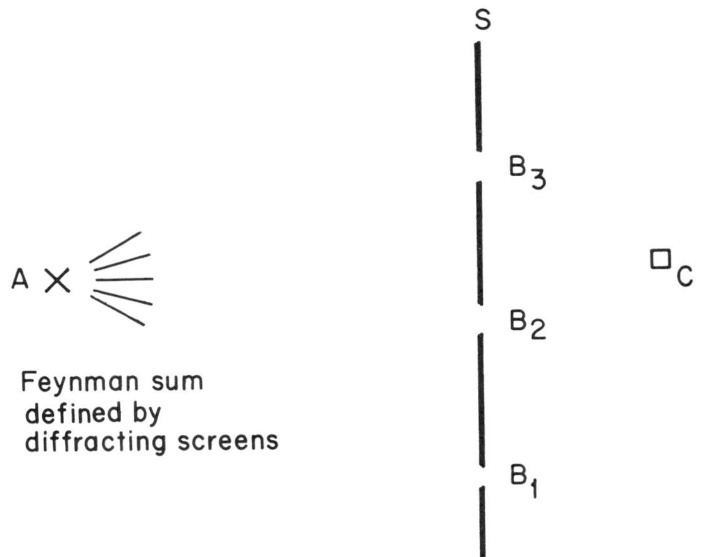

Figure 9.2

Let A be a given source of electrons. Let S be a screen with many openings B_1 B_2 ··· . Let C be a detector of those particles which have progressed from A through the openings in the screen.

If the initial momentum at A as well as the initial position is given, then the classical motion is perfectly determined and the electron will strike the screen at a definite point, which may or may not be an opening (B_k). However, if A is a source of electrons with some distribution in momentum, then there is a certain probability, according to classical theory, of reaching C. This probability is

$$P(AC) = \sum_k P(AB_kC) \tag{9.2}$$

where $P(AB_kC)$ is the probability of passing through the kth hole, since the possibilities B_k are mutually exclusive.

On the other hand, according to the superposition principle of quantum theory, the probability amplitude of arriving at C will be

$$\psi(AC) = \sum_k \psi(AB_kC) \tag{9.3}$$

where $\psi(AB_kC)$ is the probability amplitude associated with the kth opening. The probability is then

$$P(AC) = |\psi(AC)|^2 \quad . \tag{9.4}$$

If one now imagines that the space between A and C is filled with many perforated screens, one may picture the Feynman sum as the limiting expression for

$$\psi(AC) = \sum_{k\alpha} \psi(AB_{k\alpha}C) \quad , \tag{9.3a}$$

where (k,α) is the kth opening in the α screen, when the number of holes and number of screens increases indefinitely. If one wished to base quantum mechanics entirely on the Feynman principle, one would reverse this line of thought and describe the above experiments with the aid of the path integral.

(a) Extraction of Classical Action from Feynman Sum

Let us rewrite (9.1) as follows

$$\langle q't'|qt\rangle = \sum_{[q]} \exp\left(\frac{i}{\hbar} S[q]\right) \tag{9.5}$$

where [q] indicates that the sum is to be taken over all paths. Let $\overset{\circ}{q}_k(t)$ be the classical motion and let

$$q_k(t) = \overset{\circ}{q}_k(t) + u_k(t) \quad . \tag{9.6}$$

Then

$$S[q] = S[\overset{\circ}{q} + u]$$

$$= S[\overset{\circ}{q}] + \int_{t_1}^{t_2} \sum_k \left[\frac{\partial L}{\partial q_k} - \frac{d}{dt}\left(\frac{\partial L}{\partial \dot{q}_k}\right)\right]_0 u_k(t)\, dt + \Delta[u,\overset{\circ}{q}] \tag{9.7}$$

by (1.10) of this chapter. Since $\overset{\circ}{q}_k(t)$ by definition satisfies the equation of motion,

$$S[q] = S[\overset{\circ}{q}] + \Delta[u] \tag{9.8}$$

where the remainder $\Delta(u)$ depends on second and higher order variations and also on the unvaried path in general. Then

$$\langle q't'|qt\rangle = \exp\left[\frac{i}{\hbar} S(\overset{\circ}{q})\right] \sum_{[u]} \exp\left[\frac{i}{\hbar}\Delta[u]\right] \tag{9.9}$$

$$= \exp\left[\frac{i}{\hbar} S(\overset{\circ}{q})\right] K(t', t) \quad . \tag{9.9a}$$

The sum over paths [u] is an integral over the function space of the paths (path space), and may be described as follows.

Let

$$u(t) = \sum C_m V_m(t) \tag{9.10}$$

where V_m is a complete orthogonal set. Since the varied paths agree at the endpoints, choose the orthogonal set such that

$$V_m(t_1) = V_m(t_2) = 0 \qquad . \tag{9.11}$$

Then after carrying out the integration over t that is implied by Δ, one obtains

$$\Delta[u] = \Delta(C_1, C_2, \cdots) \qquad . \tag{9.12}$$

The sum over paths is now a sum over all choices of the coefficients (C_i). Therefore

$$\sum_{[u]} \exp\left[\frac{i}{\hbar}\Delta[u]\right] = \sum_{C_1C_2\cdots} \exp\left[\frac{i}{\hbar}\Delta(C_1C_2\cdots)\right] W(C_1C_2\cdots) \tag{9.13}$$

where $W(C_1C_2\cdots)$ is the weight associated with the point $(C_1C_2\cdots)$ in path space.[10] We shall illustrate the method by discussing the harmonic oscillator.

(b) The Harmonic Oscillator

In this case the Lagrangian is

$$L = \frac{1}{2}[\dot{q}^2 - \omega^2 q^2) \qquad . \tag{9.14}$$

The variation of the action [Eq. (9.8)] is

$$\Delta(u) = \int_{t_1}^{t_2} \frac{1}{2}(\dot{u}^2 - \omega^2 u^2)dt \qquad . \tag{9.15}$$

The expansion of u(t) is

$$u(t) = \sum_1^N C_k \sin \omega_k t \tag{9.16}$$

where

$$\omega_k T = k\pi \qquad k = 1,2,\cdots \tag{9.17}$$

so that

$$u(0) = u(T) = 0 \quad .$$

Then

$$\Delta(u) = \frac{T}{2} \sum_1^N (\omega_k{}^2 - \omega^2)\, C_k{}^2 \quad . \tag{9.18}$$

Define

$$K_N = \sum_{[u]} \exp\left[\frac{i}{\hbar}\,\Delta(u)\right]$$

$$= \sum_{C_1 \cdots C_N} \exp\left[\frac{iT}{2\hbar} \sum_1^N (\omega_k{}^2 - \omega^2)\, C_k{}^2\right] W(C_1 \cdots C_N) \tag{9.19}$$

as in (9.13). Here $(C_1 \cdots C_N)$ represents a single path and the sum over paths is then the integral over all points $(C_1 \cdots C_N)$ of "path space." This integral is

$$K_N = \int_{-\infty}^{+\infty} \cdots \int_{-\infty}^{+\infty} dC_1 \cdots dC_N \exp\left[\frac{i}{\hbar} S(C_1)\right] \cdots$$

$$\times \exp\left[\frac{i}{\hbar} S(C_N)\right] W(C_1 \cdots C_N) \tag{9.20}$$

where

$$S(C_k) = \frac{T}{2} (\omega_k{}^2 - \omega^2) C_k{}^2 \quad . \tag{9.20a}$$

Assume

$$W(C_1 \cdots C_N) = \prod_1^N W_k \quad , \tag{9.21}$$

and assume that all parts of path space are equally weighted so that the W_k are independent of C_k. Then

$$K_N = \prod_{k=1}^N W_k \int_{-\infty}^{+\infty} dC_k \exp\left[\frac{iT}{2\hbar} (\omega_k{}^2 - \omega^2) C_k{}^2\right]$$

$$= \prod_{k=1}^N W_k \left[\frac{2\hbar}{(\omega_k{}^2 - \omega^2)T}\right]^{1/2} \int_{-\infty}^{+\infty} \exp[ia_k{}^2]\, da_k \tag{9.22}$$

where

$$a_k{}^2 = \frac{T}{2\hbar} (\omega_k{}^2 - \omega^2) C_k{}^2 \quad . \tag{9.23}$$

Therefore

$$K_N \sim \left(\frac{1}{T^{1/2}}\right)^N \prod_{k=1}^{N} \frac{W_k}{\omega_k} \frac{1}{[1 - (\omega^2/\omega_k^2)]^{1/2}} \quad . \tag{9.24}$$

The choice of W_k is not important here since (9.24) may be factored as follows:

$$\prod_k (W_k/\omega_k)(1 - \omega^2/\omega_k^2)^{-1/2} = \prod_k (W_k/\omega_k) \prod_k (1 - \omega^2/\omega_k^2)^{-1/2}$$

where the first product on the right is independent of ω and may be ignored if we temporarily ignore the overall normalization. But

$$\left(\frac{z}{\sin z}\right)^{1/2} = \prod_k \left(1 - \frac{z^2}{k^2\pi^2}\right)^{-1/2} \quad . \tag{9.27}$$

Therefore

$$K = \lim_{N\to\infty} K_N = C(T) \left(\frac{\omega T}{\sin \omega T}\right)^{1/2} \tag{9.28}$$

where $C(T)$ depends on the weights W_k and is so far undetermined. Finally by (9.9)

$$\langle q't'|qt\rangle = C(T) \left(\frac{\omega T}{\sin \omega T}\right)^{1/2} \exp[iS(\overset{\circ}{q})/\hbar] \quad . \tag{9.28a}$$

The normalization may now be obtained from the composition

law:

$$\langle q''t''|qt\rangle = \int \langle q''t''|q't'\rangle \langle q't'|qt\rangle \, dq' \tag{9.29}$$

with the result

$$\langle q't'|qt\rangle = \left(\frac{m\omega}{2\pi \, i\hbar \, \sin \omega T}\right)^{1/2} e^{(i/h)\mathring{S}} \quad . \tag{9.30}$$

Notice that the integration over paths determines the dependence of $\langle q't'|qt\rangle$ on ω but that the unitary composition law (9.29) has been used to determine the dependence on $T = t'-t$. The result (9.30) may be compared with the alternative calculation based on Eq. (8.13).

Note that for the limiting case of a free particle ($\omega = 0$)

$$\langle q't'|qt\rangle = \frac{A}{T^{1/2}} \exp\left[\frac{i}{2\hbar} \frac{x^2}{T}\right] \quad . \tag{9.31}$$

The normalization may therefore also be obtained from the known value of A for a free particle.

If the correct modulus is introduced into the Feynman sum at the beginning of the argument in (9.1), then it is not necessary to appeal to (9.29) or (9.31) [as is shown in note (10) at the end of this chapter].

(c) Matrix Elements[11]

The matrix element of an operator $A(\underset{\sim}{q}_m)$ between the initial and final states is

$$\langle \underset{\sim}{q}_n | A(\underset{\sim}{q}_m) | \underset{\sim}{q}_1 \rangle = \int \langle \underset{\sim}{q}_n | A(\underset{\sim}{q}_m) | \underset{\sim}{q}_m \rangle \langle \underset{\sim}{q}_m | \underset{\sim}{q}_1 \rangle \, d\underset{\sim}{q}_m$$

$$= \int A(\underset{\sim}{q}_m) \langle \underset{\sim}{q}_n | \underset{\sim}{q}_m \rangle \langle \underset{\sim}{q}_m | \underset{\sim}{q}_1 \rangle \, d\underset{\sim}{q}_m \qquad (9.32)$$

where $\underset{\sim}{q}_1$, $\underset{\sim}{q}_m$, and $\underset{\sim}{q}_n$ are the complete set of coordinates taken at times t_1, t_m, and t_n. (Notice that there is no sum on m in this equation.) By substituting for the transformation functions on the right one may show that

$$\langle \underset{\sim}{q}_n | A(\underset{\sim}{q}_m) | \underset{\sim}{q}_1 \rangle = \sum_p A(\underset{\sim}{q}_m, p) \, e^{(i/\hbar) S_p} \quad , \qquad (9.33)$$

and one may also generalize (9.33) to the following:

$$\langle 2 | A | 1 \rangle = \sum_p A_p \, e^{(i/\hbar) S_p} \qquad (9.34)$$

where A_p is in general a function of the path p, such as a time ordered product or a functional integral, that is weighted by the phase factor $\exp[(i/\hbar) S_p]$. Equation (9.34) of course contains the special case $\langle 2 | 1 \rangle$ when A = 1.

3.10 THE SCHWINGER ACTION PRINCIPLE[12]

According to the Feynman principle, one has a complete solution to the quantal problem, if the functional integral defined by the sum over all paths can be carried out. The problem of actually carrying out the functional integration is, however, much more difficult than calculating the classical action by integration according to the definition

$$S = \int L \, dt \quad .$$

If this classical integration can be accomplished for all comparison paths, then one can in principle select the extremum to obtain a solution of the dynamical problem; and furthermore the action integral for the actual motion is a solution of the Hamilton-Jacobi equation. Ordinarily it is not possible to carry out even this classical integration and of course it is even qualitatively much more difficult to carry out the corresponding functional integration.

The Schwinger action principle avoids the functional integration since it does not deal directly with $\langle 2|1\rangle$ but rather with $\delta\langle 2|1\rangle$ where the variation may be real or virtual. In this respect the Schwinger principle is precisely analogous to classical variational principles, and consequently leads to differential equations rather than to the solution of these equations.

Schwinger defines an operator δW_{21} as follows:

$$\delta \langle 2|1\rangle = \frac{i}{\hbar} \langle 2|\delta W_{21}|1\rangle \qquad . \tag{10.1}$$

The properties of W_{21} are determined by the transformation function which in turn has the following properties:

(a) Composition law:

$$\langle a'|c'\rangle = \sum_{b'} \langle a'|b'\rangle \langle b'|c'\rangle \qquad . \tag{10.2}$$

(b) Unitarity:

$$\langle a'|c'\rangle = \langle c'|a'\rangle^* \qquad . \tag{10.3}$$

It follows from (10.2) that

$$\begin{aligned}\delta\langle 3'|1'\rangle &= \sum [(\delta\langle 3'|2'\rangle) \langle 2'|1'\rangle + \langle 3'|2'\rangle (\delta\langle 2'|1'\rangle)] \\ &= \frac{i}{\hbar} \sum [\langle 3'|\delta W_{32}|2'\rangle\langle 2'|1'\rangle + \langle 3'|2'\rangle\langle 2'|\delta W_{21}|1'\rangle] \\ &= \frac{i}{\hbar} \langle 3'|\delta W_{32} + \delta W_{21}|1'\rangle \qquad .\end{aligned}$$

In this abbreviated notation the index 2 refers to time t_2 and 2' means the quantum numbers like q'_2 chosen at t_2. Therefore

$$\delta W_{31} = \delta W_{32} + \delta W_{21} \qquad . \tag{10.4}$$

From (10.4)

$$\delta W_{11} = 0 \tag{10.5a}$$

$$\delta W_{12} = -\,\delta W_{21} \quad . \tag{10.5b}$$

It follows from (10.3) that

$$\delta\langle 1'|2'\rangle = \delta\langle 2'|1'\rangle^*$$

$$= -\frac{i}{\hbar}\langle 2'|\delta W_{12}|1'\rangle^* \quad .$$

By (10.5b)

$$\langle 1'|\delta W_{12}|2'\rangle = \langle 2'|\delta W_{12}|1'\rangle^* \quad . \tag{10.6}$$

Hence δW_{12} is hermitian.

The properties (10.4) and (10.6) now follow if one assumes the representation

$$W_{21} = \int_{t_1}^{t_2} L\, dt \tag{10.7}$$

where L is hermitian:

$$L = L^+ \quad . \tag{10.7a}$$

Then (10.1) may be restated:

$$\delta\langle 2'|1'\rangle = \frac{i}{\hbar}\langle 2'|\delta\int_{t_1}^{t_2} L\,dt|1'\rangle \tag{10.8}$$

The dynamical system is now characterized by L, which is defined to be the quantum mechanical Lagrangian. In the correspondence limit L may be interpreted in terms of the classical Lagrangian.

We may now turn the argument around and regard (10.8) as a condition that determines the transformation function $\langle 2'|1'\rangle$ in terms of a given Lagrangian. In this form it is known as the Schwinger action principle. Let us now examine the simplest consequences of this principle.

(a) Hamilton's Principle

Let us first apply (10.8) in the same way as the classical principle of Hamilton. Then initial and final states are not varied

$$\delta|1\rangle = \delta\langle 2| = 0 \tag{10.9}$$

where the primes have been dropped. Therefore

$$\langle 2|\delta\int_{t_1}^{t_2} L\,dt|1\rangle = 0 \quad . \tag{10.10}$$

Since 1 and 2 are arbitrary, we have the operator variational principle

$$\delta \int_{t_1}^{t_2} L \, dt = 0 \quad . \tag{10.11}$$

The Euler-Lagrange and the Hamiltonian equations following from (10.11) are now the quantal operator equations of motion.[12] In carrying out the variation, one must of course preserve order, but one may choose variations which commute with all operators in a known way.

(b) Schrödinger Equation

The Schrödinger equation corresponds to the Hamilton-Jacobi equation in the correspondence limit. Since the latter arises from an endpoint variation in the classical theory, one should expect the Schrödinger equation to arise from an end-point variation in the Schwinger variational principle. Then assume

$$\delta|1\rangle = 0 \qquad \delta|2\rangle \neq 0$$
$$\delta q_1 = \delta t_1 = 0 \qquad \delta q_2 \neq 0 , \quad \delta t_2 \neq 0 \quad .$$

Next defining the Hamiltonian and conjugate momenta as in the classical case but with due regard for order of factors, one may show

$$\delta\langle 2|1\rangle = \frac{i}{\hbar} \langle 2|-\delta t_2 H + \delta q_2 p_2|1\rangle \tag{10.12}$$

and, restoring the primes,

$$-\frac{\hbar}{i} \frac{\delta\langle 2'|1'\rangle}{\delta t_2} = \langle 2'|H|1'\rangle \tag{10.13}$$

$$\frac{\hbar}{i} \frac{\delta\langle 2'|1'\rangle}{\delta q'_2} = \langle 2'|p_2|1'\rangle \quad . \tag{10.14}$$

Equation (10.13) is the Schrödinger equation for the kernel or transformation function $\langle 2'|1'\rangle$ according to (8.12a). Equation (10.14) gives the representation of the momentum operator also in terms of the transformation function $\langle 2|1\rangle$ according to (4.12) of Chapter 2.

The proper commutation rules between p and q are implied by (10.14). [Compare with (4.12) of Chapter 2.]

Equations (10.13) and (10.14) arise in just the same general way as the classical equations (2.12) and (2.11) of this chapter.

The Schrödinger equation as well as the canonical commutation rules may in this way be derived from the action principle (10.8). Since the equivalence of the Schrödinger and Heisenberg pictures has already been shown, one sees that the action principle must also lead to the Heisenberg picture.

(c) Unitary Transformations

Let

$$|1\rangle' = U|1\rangle \quad . \tag{10.15}$$

If U is close to the identity, we can write

$$U = 1 + \frac{1}{i\hbar} G \tag{10.15a}$$

where G is infinitesimal. Then G is called the hermitian generator and

$$\delta|\ \rangle = \frac{1}{i\hbar} G|\ \rangle \tag{10.16a}$$

and

$$\delta(\langle\ |) = -\frac{1}{i\hbar}\langle\ |G \quad . \tag{10.16b}$$

The corresponding variation in the transformation function is

$$\delta\langle 2|1\rangle = \frac{1}{i\hbar}\langle 2|-G_2 + G_1|1\rangle \tag{10.17}$$

if one supposes different generators to work at times t_1 and t_2.

By (10.17) and (10.8)

$$\langle 2|\delta\int_{t_1}^{t_2} L\ dt|1\rangle = \langle 2|G_2 - G_1|1\rangle$$

or as an operator equation

$$G_2 - G_1 = \delta \int_{t_1}^{t_2} L\, dt \quad . \tag{10.18}$$

It follows that one may choose

$$\delta L = \frac{dG}{dt}$$

or

$$L' = L + \frac{dG}{dt} \quad . \tag{10.19}$$

Therefore the addition of dG/dt to the Lagrangian is equivalent to making unitary transformations on the initial and final state functions. In other words there is an arbitrariness in the Lagrangian that corresponds exactly to the freedom of making unitary transformations on the initial and final state vectors.

We introduced classical contact transformations in paragraph (3.4) in just this same way, namely by adding a perfect differential to the integrand of the action integral. Therefore, we now have a direct way of looking at the equivalence of unitary transformations and contact transformations.

3.11 FORMAL INTEGRATION OF ACTION PRINCIPLE

Consider

$$\langle b'|O|a'\rangle \tag{11.1}$$

where O is some function of the operators A and B. This matrix element is immediately evaluated in the following cases:

$$\langle b'|f(A)|a'\rangle = f(a')\,\langle b'|a'\rangle \tag{11.2}$$

$$\langle b'|g(B)|a'\rangle = g(b')\,\langle b'|a'\rangle \tag{11.3}$$

$$\langle b'|g(B)\,f(A)|a'\rangle = g(b')\,f(a')\,\langle b'|a'\rangle \tag{11.4}$$

but $\langle b'|AB|a'\rangle$ cannot be readily evaluated unless the commutator (A,B) is simple. If

$$(A,B) = i\hbar$$

then

$$AB = BA + i\hbar \tag{11.5}$$

$$\begin{aligned}\langle b'|AB|a'\rangle &= \langle b'|BA + i\hbar|a'\rangle \\ &= [b'a' + i\hbar]\,\langle b'|a'\rangle \quad .\end{aligned} \tag{11.6}$$

Any function O(A,B) can be similarly handled. Let

$$O(A,B) = \mathcal{O}(A,B) \tag{11.7}$$

where $\mathcal{O}(A,B)$ is obtained from $O(A,B)$ by repeated application of a relation like (11.5) so that every term in $\mathcal{O}(A,B)$ is of form

$$B^n A^m \quad .$$

Then $\mathcal{O}$ is called well-ordered[12] and

$$\langle b'|O|a'\rangle = \langle b'|\mathcal{O}|a'\rangle = \mathcal{O}'\langle b'|a'\rangle \quad . \tag{11.8}$$

This procedure may now be used to obtain a formal solution of the Schwinger condition for the transformation function.

Let

$$\delta\langle 2'|1'\rangle = \frac{i}{\hbar}\langle 2'|\delta W_{21}|1'\rangle \tag{11.9}$$

where 1' and 2' stand for the set of quantum numbers which label initial and final states.

Let

$$\delta W_{21} = \delta \mathcal{W}_{21} \tag{11.10}$$

where $\delta\mathcal{W}$ is the well-ordered form of δW. Then

$$\delta\langle 2'|1'\rangle = \frac{i}{\hbar}\langle 2'|\delta\mathcal{W}_{21}|1'\rangle = \frac{i}{\hbar}\,\delta\mathcal{W}_{21}'\,\langle 2'|1'\rangle$$

$$\delta\,\ell n\,\langle 2'|1'\rangle = \frac{i}{\hbar}\,\delta\mathcal{W}_{21}'$$

$$\langle 2'|1'\rangle = \exp\left(\frac{i}{\hbar}\,\mathcal{W}_{21}'\right) \quad . \tag{11.11}$$

This result resembles the Feynman formula (9.1) but contains the eigenvalue of a well-ordered exponent instead of a sum over paths. In both cases, however, the exponent is the action integral. On the classical path the function S in (9.1) satisfies the Hamilton-Jacobi equation. It will now be shown that $\mathcal{W}$ satisfies an operator Hamilton-Jacobi equation.

3.12 THE HAMILTON-JACOBI OPERATOR EQUATION[12,13]

According to (10.13) and (10.14) of this chapter, we find with the aid of (11.11)

$$\begin{aligned} \langle 2'|H|1'\rangle &= -\frac{\hbar}{i}\frac{\partial}{\partial t_2}\langle 2'|1'\rangle \\ &= -\frac{\hbar}{i}\frac{\partial}{\partial t_2}\exp\left(\frac{i}{\hbar}\mathcal{W}_{21}'\right) \end{aligned} \tag{12.1}$$

$$\begin{aligned} \langle 2'|p_2|1'\rangle &= \frac{\hbar}{i}\frac{\partial}{\partial q_2}\langle 2'|1'\rangle \\ &= \frac{\hbar}{i}\frac{\partial}{\partial q_2'}\exp\left(\frac{i}{\hbar}\mathcal{W}_{21}'\right) \quad . \end{aligned} \tag{12.2}$$

Equation (12.2) becomes

$$\langle 2'|p_2|1'\rangle = \frac{\partial \mathcal{W}_{21}'}{\partial q_2'}\langle 2'|1'\rangle = \langle 2'|\,\frac{\partial \mathcal{W}_{21}}{\partial q_2}\,|1'\rangle \tag{12.3a}$$

or

$$p_2 = \frac{\partial \mathcal{W}}{\partial q_2} \quad . \tag{12.3b}$$

Similarly (12.1) becomes

$$\langle 2'|H|1'\rangle = -\frac{\partial \mathcal{W}_{21}{}'}{\partial t_2} \langle 2'|1'\rangle = -\langle 2'|\frac{\partial \mathcal{W}_{21}}{\partial t_2}|1'\rangle \tag{12.4a}$$

and

$$H + \frac{\partial \mathcal{W}}{\partial t} = 0 \tag{12.4b}$$

or

$$H\left(\frac{\partial \mathcal{W}}{\partial q} \cdots q \cdots\right) + \frac{\partial \mathcal{W}}{\partial t} = 0 \quad . \tag{12.5}$$

This last equation (12.5) is the operator Hamilton-Jacobi equation. It arises from a variation of the final state just as the Schrödinger equation does.

The correspondence limit of $\mathcal{W}_{21}{}'$ is S and the correspondence limit of (12.5), as well as of the Schrödinger equation, is the classical Hamilton-Jacobi equation.

Example (The Harmonic Oscillator)

The classical Hamilton-Jacobi equation is

$$\frac{1}{2}\left(\frac{\partial S}{\partial q}\right)^2 + \frac{\omega^2}{2} q^2 + \frac{\partial S}{\partial t} = 0 \tag{12.6}$$

with the solution

$$S = \frac{\omega}{2 \sin \omega t} [(q^2 + q_o^{\ 2}) \cos \omega t - 2q_o q] \quad . \tag{12.7}$$

The corresponding operator equation is

$$\frac{1}{2} \left(\frac{\partial W}{\partial q}\right)^2 + \frac{\omega^2}{2} q^2 + \frac{\partial W}{\partial t} = 0 \quad . \tag{12.8}$$

Let us make the ansatz

$$W = \overleftarrow{S} + \phi(t) \tag{12.9}$$

where $\overleftarrow{S}$ means that q is written to the left of q_o in (12.7):

$$\overleftarrow{S} = \frac{\omega}{2 \sin \omega t} [(q^2 + q_o^{\ 2}) \cos \omega t - 2qq_o] \tag{12.9a}$$

and where $\phi(t)$ commutes with all operators and must vanish in the classical limit. Then the momentum operator is

$$p = \frac{\partial W}{\partial q} = \frac{\partial \overleftarrow{S}}{\partial q} = \frac{\omega}{\sin \omega t} (q \cos \omega t - q_o) \tag{12.10}$$

and the commutator of q and q_o may be obtained from

$$(q,p) = i\hbar \quad ,$$

which now becomes, because of (12.10),

$$(q_o,q) = i\hbar \left(\frac{\sin \omega t}{\omega}\right) \quad . \tag{12.11}$$

According to (12.11), q_o and q commute only at the same time. From (12.10) one also finds

$$\left(\frac{\partial W}{\partial q}\right)^2 = \left(\frac{\omega}{\sin \omega t}\right)^2 [q^2 \cos^2 \omega t - (q_o q + q q_o) \cos \omega t + q_o^2) \tag{12.12}$$

and by (12.11) it is possible to express $(\partial W/\partial q)^2$ also in the ordered form:

$$\left(\frac{\partial W}{\partial q}\right)^2 = \left(\frac{\omega}{\sin \omega t}\right)^2 (q^2 \cos^2 \omega t - 2q\, q_o \cos \omega t + q_o^2) + \frac{\hbar}{i}\left(\frac{\omega}{\sin \omega t}\right) \cos \omega t \quad . \tag{12.13}$$

Again from (12.9) one finds

$$\frac{\partial W}{\partial t} = \frac{1}{2}\left(\frac{\omega}{\sin \omega t}\right)^2 (2q\, q_o \cos \omega t - q^2 - q_o^2) + \phi' \tag{12.14}$$

and therefore from the operator Hamilton-Jacobi equation (12.8), and (12.13) and (12.14) one obtains

$$\frac{1}{2}\left(\frac{\omega}{\sin \omega t}\right)^2 (q^2 \cos^2 \omega t - 2q\, q_o \cos \omega t + q_o^2) - \frac{i\hbar}{2}\omega \cot \omega t + \frac{\omega^2 q^2}{2} + \frac{1}{2}\left(\frac{\omega}{\sin \omega t}\right)^2 (2q\, q_o \cos \omega t - q^2 - q_o^2) + \phi' = 0$$

or

$$\phi' - \frac{i\hbar}{2} \omega \cot \omega t = 0 \tag{12.15}$$

with the solution

$$\phi(t) = \frac{i\hbar}{2} \ln (A \sin \omega t) \quad . \tag{12.16}$$

Then by (11.11) and (12.9)

$$\langle q_2 t_2 | q_1 t_1 \rangle = \exp\left(\frac{i}{\hbar} \phi(t)\right) \exp\left(\frac{i}{\hbar} S\right)$$

where S is simply the classical action. By (12.16)

$$\exp\left(\frac{i}{\hbar} \phi(t)\right) = \exp\left(- \frac{1}{2} \ln(A \sin \omega t)\right) = \left(\frac{1}{A \sin \omega t}\right)^{1/2}$$

or

$$\langle q_2 t_2 | q_1 t_1 \rangle \sim \left(\frac{1}{\sin \omega t}\right)^{1/2} \exp\left(\frac{i}{\hbar} S\right) \tag{12.17}$$

in agreement with (9.30) obtained by summing over paths.

We summarize this section by presenting the Feynman and Schwinger recipes in the following form

$$\langle 2 | 1 \rangle = \sum_p e^{\frac{i}{\hbar} S_p} \tag{12.18a}$$

$$\langle 2 | 1 \rangle = e^{\frac{i}{\hbar} W_{21}'} \tag{12.18b}$$

where S_p satisfies the classical Hamilton-Jacobi equation on the classical path and where $\mathcal{W}_{21}$ satisfies the operator Hamilton-Jacobi equation.

Finally we note that the Schwinger principle may be obtained by variation of the Feynman statement since

$$\delta\langle 2|1\rangle = \frac{i}{\hbar}\sum_p \delta S_p\, e^{iS_p/\hbar} = \frac{i}{\hbar}\langle 2|\delta S|1\rangle$$

by (9.34), and therefore all the consequences of the Schwinger principle also follow from the Feynman formulation.

3.13 BROWNIAN MOTION AND QUANTUM MECHANICAL PROPAGATION

The intuitive approach (described in Section 9) to the Feynman sum over paths may be developed further. There the probability amplitude for going from (x_1,t_1) to (x_2,t_2) is expressed in (9.1) as

$$\langle x_2t_2|x_1t_1\rangle = \sum_p e^{(i/\hbar)S_p(2,1)} \tag{13.1}$$

where the sum is over all spacetime paths (p) leading from (x_1t_1) to (x_2t_2). The corresponding classical expression for the probability is

$$P(x_2t_2|x_1t_1) = \sum_p P_p(x_2t_2|x_1t_1) \tag{13.2}$$

where $P_p(x_2t_2|x_1t_1)$ is the probability of classically following the particular path, p, from (x_1t_1) to (x_2t_2). Here $P(x_2t_2|x_1t_1)$ is the conditional probability that the particle is found at x_2 at time t_2 if it is given at x_1 at the earlier time t_1.

If the particle follows a random walk, then it may be shown that[14]

$$P(x_2t_2|x_1t_1) = \frac{1}{(4\pi Dt)^{3/2}} \exp[-(x_2-x_1)^2/4D(t_2-t_1)] \tag{13.3}$$

where

$$t = t_2 - t_1 \tag{13.3a}$$

$$D = \frac{1}{2} n\ell^2 \quad . \tag{13.3b}$$

Here n is the number of displacements per unit time and ℓ is the length of a step. Then it is also easy to show that $P(x_2t_2|x_1t_1)$ satisfies the diffusion equation

$$\frac{\partial P}{\partial t} = D\nabla^2 P \tag{13.4}$$

while $\langle x_2t_2|x_1t_1\rangle$ satisfies the Schrödinger equation

$$\frac{\partial \psi}{\partial t} = \frac{i\hbar}{2m} \nabla^2\psi \quad . \tag{13.5}$$

If the diffusion coefficient D is replaced by $i\hbar/2m$, then the probability of having moved from an initial to a final point by a random walk, or a Brownian motion, goes over into the probability amplitude for having traversed the same interval by quantum mechanics.

There is also a functional integral, the Wiener integral,[15] that is related to the classical problem in the same way that the Feynman integral is related to the quantal problem. These two functional integrals may be compared if one defines the Feynman amplitude by first constructing the following expression for a broken path

$$\langle x_n t_n | x_1 t_1 \rangle = \langle x_n t_n | x_{n-1} t_{n-1} \rangle \langle x_{n-1} t_{n-1} | \rangle \cdots \langle x_2 t_2 | x_1 t_1 \rangle \tag{13.6}$$

and summing over all such paths leading from the given initial to the given final point. Then in the same way one may define the Wiener functional integral by constructing the corresponding probability for the same broken path

$$P(x_n t_n | x_1 t_1) = \\ = P(x_n t_n | x_{n-1} t_{n-1}) \, P(x_{n-1} t_{n-1} |) \cdots P(x_2 t_2 | x_1 t_1) \tag{13.7}$$

and integrating over this class of paths. The integrations in the two cases depend on the corresponding composition laws.

For unitary transformations the composition law is

$$\langle x_3 t_3 | x_1 t_1 \rangle = \int \langle x_3 t_3 | x_2 t_2 \rangle \langle x_2 t_2 | x_1 t_1 \rangle \, dx_2 \quad . \qquad (13.8)$$

The corresponding law for probabilities is called the Smoluchowski equation

$$P(x_3 t_3 | x_1 t_1) = \int P(x_3 t_3 | x_2 t_2) \, P(x_2 t_2 | x_1 t_1) \, dx_2 . \qquad (13.9)$$

NOTES ON CHAPTER 3

1. Kinetic Focus

In phase space there is only one dynamical trajectory passing through each point, but in configuration space there may be a pencil of paths which focuses at some second point (see Figure).

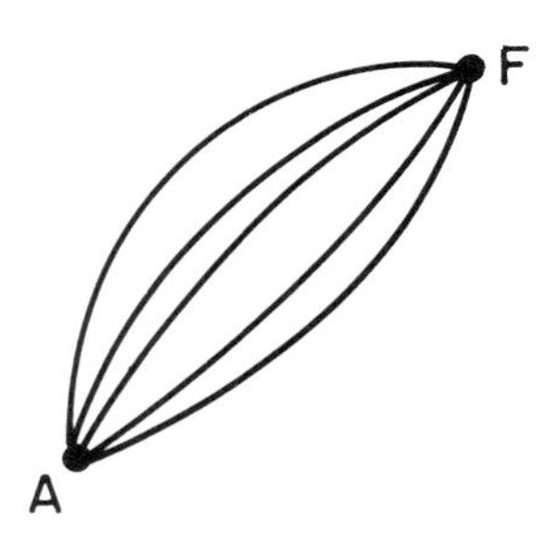

Kinetic Focus

If A is the initial point then the limiting position of F, as the angle of the pencil emerging from A approaches zero, is a kinetic focus. One may show that if the endpoint lies between A and F, then S is a minimum. Otherwise S is neither a maximum nor a minimum. See E. T. Whittaker, op. cit. in Chapter 1.

2. The coordinate q_k is said to be "ignorable" if it does not appear in the Lagrangian. Therefore $\delta S = 0$ when q_k is varied. Therefore $p_k^A = p_k^B$, where p_k is the momentum conjugate to q_k and q_k is any ignorable coordinate.

The description of an N-particle system by 3N coordinates is redundant if every observable property of the system is independent of where the system is located, how it is oriented, and when the observation is made. The three coordinates locating the center of mass, and the three angles fixing the orientation are "ignorable" and the corresponding conjugate momenta are conserved. These are the conservation laws of momentum and angular momentum. Similarly the conservation law corresponding to the ignorable time is the conservation of energy. A second way to describe this situation is to say that the Hamiltonian is unchanged by a certain group of transformations, the invariance group, that in this case consists of space translations and rotations, and displacements in time. In still different language one asserts that spacetime is homogeneous and isotropic unless there is a reason for making a contrary assumption. Then one says that the laws of physics are the same at all places and at all times. The laws of physics also do not depend on the state of motion of the laboratory provided that the laboratory is an inertial frame. We have discussed all the conservation laws earlier in terms of the invariance group.

3. The situation is different in relativistic mechanics. Then the Lagrangian is an invariant although the Hamiltonian is not. (The Lagrangian of a single particle in relativity theory is simply proportional to $\int ds$, the invariant path length.)

4. For an interesting proof of the invariance of the P.B. under canonical transformation, see Landau and Lifshitz, page 145, op. cit. in Chapter 2.

5. Symplectic Invariant

One may compare

$$\delta^{mn} = \delta^{st} \frac{\partial x'^m}{\partial x^s} \frac{\partial x'^m}{\partial x^t}$$

with

$$\varepsilon^{mn} = \varepsilon^{st} \frac{\partial x'^m}{\partial x^s} \frac{\partial x'^m}{\partial x^t} \quad .$$

These equations define orthogonal and symplectic transformations respectively. The corresponding invariants are

$$A \cdot B = \delta^{mn} A_m B_n$$

and

$$[A,B] = \varepsilon^{mn} A_m B_n$$

where

$$A_m = \frac{\partial A}{\partial x^m} \quad , \qquad B_m = \frac{\partial B}{\partial x^m} \quad .$$

6. Dirac, op. cit., page 121.
7. See W. Pauli, in L. de Broglie, Physicien et Penseur, Editions Albin Michel, Paris, 1953. D. Bohm, Phys. Rev. 85, 166, 180 (1952).
8. Titchmarsh, op. cit., page 78.
9. Feynman and Hibbs, reference 2 of this chapter. See also Dirac, op. cit., for the first discussion of Lagrangian methods.
10. See preceding reference (9), page 33.

The point with coordinates $(c_1 c_2 \cdots)$ represents a path. The sum over paths is then an integration over this function space ("path space"). Such a procedure is not well defined, however, unless there is a way to assign $W(c_1 c_2 \cdots)$. Feynman's original way of writing down the sum over paths follows directly from (9.3a) and is illustrated in the figure. One divides the total time $t_B - t_A$ into small intervals $\Delta t = \varepsilon$. Then by allowing the vertices of the broken path to vary from $-\infty$ to $+\infty$ one generates an infinite set of paths. Feynman then writes for a single particle in one spatial dimension:

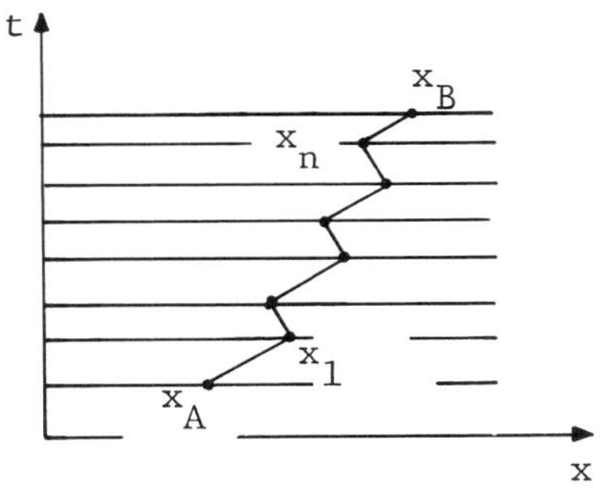

Feynman Path

$$\langle x_B t_B | x_A t_A \rangle = \lim_{\varepsilon \to 0} \frac{1}{A} \int \cdots \int e^{(i/\hbar)S} \frac{dx_1}{A} \cdots \frac{dx_n}{A} \qquad (1)$$

where n becomes infinite as $\varepsilon \to 0$. Here the $x_1 \cdots x_n$ are the intermediate values of x, while x_A and x_B are not integrated over.

Since spacetime is homogeneous in the time as well as the spatial direction, the weights associated with the different integrations are assumed all equal.

For the general system investigated in this course one may write

$$<\underset{\sim}{x}_B^{(1)}\cdots\underset{\sim}{x}_B^{(N)}t_B|\underset{\sim}{x}_A^{(1)}\cdots\underset{\sim}{x}_A^{(N)}t_A>$$

$$= \lim_{\varepsilon\to 0}\prod_{r=1}^{N}\frac{1}{A^3}\int\cdots\int e^{(i/\hbar)S}\frac{d\underset{\sim}{x}_1^{r}}{A^3}\cdots\frac{d\underset{\sim}{x}_n^{r}}{A^3} \quad . \tag{2}$$

One now comes to the problem of assigning A. In the text the normalization was determined by the composition law (9.29) for unitary transformations. It also follows from (9.29) that

$$\lim_{t'\to t} <q't'|qt> = \delta(q'-q) \tag{3}$$

and this condition is in fact sufficient for determining A, as we shall now show.

For L we shall assume

$$L = \frac{1}{2}\sum m\,\dot{\underset{\sim}{x}}_r^{\,2} - V(\underset{\sim}{x}_1\cdots\underset{\sim}{x}_N) \quad . \tag{4}$$

Then

$$L\,\Delta t = \frac{1}{2}\sum\frac{m}{\Delta t}(\Delta\underset{\sim}{x}_r)^2 - V\,\Delta t \tag{5}$$

and therefore

$$\lim_{\Delta t\to 0} <q_{t+\Delta t}\,t+\Delta t|q_t t> = \lim_{\Delta t\to 0}\prod_{r=1}^{3N}\exp\left[\frac{i}{2\hbar}\,m\,\frac{(\Delta q_r)^2}{\Delta t}\right]\frac{1}{A} \tag{6}$$

where we changed to generalized coordinates q_r for convenience in writing. It now follows from (3) that

$$\lim_{\varepsilon\to 0} \frac{\exp\left[\frac{i}{2\hbar} m \frac{(\Delta q_r)^2}{\Delta t}\right]}{A} = \delta(\Delta q_r) \quad . \tag{7}$$

Therefore

$$A = \left(\frac{2\pi \; i\hbar \; \Delta t}{m}\right)^{1/2} \quad . \tag{8}$$

It is clear from this derivation that the normalization is independent of the potential and in particular is the same normalization that one would find if there were no interaction at all. This same feature is also apparent in the text where the normalization of the oscillator can be obtained by putting $\omega = 0$.

There is another formulation [C. Garrod, Rev. Mod. Phys. 38, 483 (1966)], namely:

$$\langle x_B t_B | x_A t_A \rangle =$$

$$= \lim_{\varepsilon\to 0} \int \cdots \int \exp\left[\frac{i}{\hbar} \int (p\dot{x}-H)dt\right] \frac{dp_1}{h} \cdots \frac{dp_{n+1}}{h} dx_1 \cdots dx_n \tag{9}$$

instead of (1). We have again gone back to the case of only one degree of freedom. As in the classical

Hamiltonian method one should regard p and x as independent variables where they appear in (9). Then

$$\lim_{\Delta t\to 0} \langle x,t+\Delta t|xt\rangle = \lim_{\Delta t\to 0} \int e^{\frac{i}{\hbar}(p\Delta x - H\Delta t)} \frac{dp}{h} \tag{10}$$

$$= \lim_{\Delta t\to 0} \int e^{\frac{i}{\hbar}(p\Delta x - \frac{p^2}{2m}\Delta t)} \frac{dp}{h} \tag{11}$$

$$= \lim_{\Delta t\to 0} \frac{1}{A} e^{\frac{i}{\hbar}\frac{m}{2}\frac{(\Delta x)^2}{\Delta t}} \tag{12}$$

where the A that appears in (12) comes from the integration of (11) and agrees with (8). Therefore the phase space integration in (9) leads exactly to the correct normalization.

11. See, for example, Yourgrau and Mandelstam, reference 4 of this chapter.

12. For details of this development see Schwinger, reference 3 of this chapter.

13. Dirac, op. cit.

14. See, for example, S. Chandrasekhar, Rev. Mod. Phys. 15, 1 (1943).

15. L. M. Gel'fand and A. M. Yaglom, J. Math. Phys. 1, 48 (1960).

BIBLIOGRAPHY FOR CHAPTER 3

General Works

In addition to references given after previous chapters, the following are suggested:

1. C. Lanczos, The Variational Principles of Mechanics, Univ. of Toronto Press, Toronto (1949).
2. Feynman and Hibbs, Quantum Mechanics and Path Integrals, McGraw-Hill.
3. Julian Schwinger, Quantum Kinematics and Dynamics, Benjamin (1970).
4. M. Yourgrau and S. Mandelstam, Variational Principles in Dynamics and Quantum Theory, Saunders (1968), p. 133.
5. G. Rosen, Formulations of Classical and Quantum Dynamical Theory, Academic Press (1969).
6. A. Katz, Classical Mechanics, Quantum Mechanics, Field Theory, Academic Press (1965).

PROBLEMS

1. The equations of an infinitesimal contact transformation are

$$Q_k = q_k + \varepsilon \frac{\partial G}{\partial P_\mu} \quad , \qquad P_k = p_k - \varepsilon \frac{\partial G}{\partial q_k} \quad ,$$

where $G(q,P)$ is the generator. Apply to rotations and Galilean transformations.

2. Find the contact transformation generated by

$$S(q,Q,T) = \frac{1}{2T} \sum_s (q_s - Q_s)^2 \quad .$$

Find the limit as $q_s \to Q_s$ and $T \to 0$.

3. Show that Poisson brackets are invariant under the following canonical transformation.

$$p_k = \frac{\partial F}{\partial q_k}(q,P,t) \quad , \qquad Q_k = \frac{\partial F}{\partial P_k}(q,P,t) \quad .$$

4. The Lagrangian of an harmonic oscillator is

$$L = \frac{1}{2}(\dot{q}^2 - q^2) \quad .$$

Write Hamiltonian and Hamilton-Jacobi equation. Find $\int L\, dt$ and show that it is a solution of the Hamilton-Jacobi equation.

5. Find $\langle x't' | xt \rangle$ for harmonic oscillator with the aid of (8.13). Compare result with (9.30) by making use of action computed in preceding exercise.

6. Calculate $\langle x't' | xt \rangle$ for a free particle by (8.8).

7. Prove (7.11) of this chapter.

8. Show that the P.B. of the position vectors at different times $x_\alpha(t)$ and $x_\alpha(t_o)$ is

$$[x_\alpha(t), x_\beta(t_o)] = -\frac{\partial}{\partial P_\beta}\int_{t_o}^{t} \frac{\partial H}{\partial p_\alpha}\, dt' \quad .$$

Evaluate this expression for a harmonic oscillator.

9. Lagrange Brackets

Define $\varepsilon_{k\ell} = \begin{pmatrix} 0 & -1 \\ 1 & 0 \end{pmatrix}$; $\varepsilon^{k\ell} = \begin{pmatrix} 0 & 1 \\ -1 & 0 \end{pmatrix}$.

Then

$$\varepsilon_{k\ell}\,\varepsilon^{\ell m} = \begin{pmatrix} 0 & -1 \\ 1 & 0 \end{pmatrix}\begin{pmatrix} 0 & 1 \\ -1 & 0 \end{pmatrix} = \begin{pmatrix} 1 & 0 \\ 0 & 1 \end{pmatrix} = \delta_k{}^m \quad .$$

Define Lagrange brackets

$$\{U_i, U_j\} = \frac{\partial x^m}{\partial U^i}\,\varepsilon_{mn}\,\frac{\partial x^n}{\partial U^j}$$

and Poisson brackets

$$[U^i, U^j] = \frac{\partial U^i}{\partial x^m}\,\varepsilon^{mn}\,\frac{\partial U^j}{\partial x^n} \quad .$$

Show that

$$\sum \{U_i, U_j\} [U^j, U^k] = \delta_i^{\ k} \quad .$$

10. Show that Lagrange brackets are invariant under canonical transformations.

11. The bilinear covariant of the differential form $\sum p_s \, dq_s$ is $\sum_s (dq_s \, \delta p_s - \delta q_s \, dp_s)$ or $\sum \varepsilon_{mn} \, dx^m \, \delta x^n$. Show that this bilinear form is invariant under canonical transformations.

12. Integral Invariants of Poincaré

Show that the following set of integrals is invariant under canonical transformations

$$J_1 = \iint \sum_i dq_i \, dp_i$$

$$J_2 = \iiiint \sum_{i \neq j} dq_i \, dp_i \, dq_j \, dp_j$$

...

$$J_n = \int \cdots \int dq_1 \, dp_1 \cdots dq_n \, dp_n \quad .$$

Notice that

$$\sum_i dq_i \, dp_i = \{U, V\} \, dU \, dV$$

where

$$\{U, V\} = \sum_i \left(\frac{\partial q_i}{\partial U} \frac{\partial p_i}{\partial V} - \frac{\partial p_i}{\partial U} \frac{\partial q_i}{\partial V} \right)$$

is the Lagrange bracket.

13. Define

$$F'(p,Q,t) = F(q,Q,t) - \sum q_i p_i \qquad \text{(a)}$$

and

$$F''(p,P,t) = F(q,Q,t) + \sum (P_i Q_i - p_i q_i) \quad . \qquad \text{(b)}$$

Show that F' and F'' define the following contact transformations:

(a)	(b)
$q_k = -\frac{\partial F'}{\partial p_k}$	$q_k = -\frac{\partial F''}{\partial p_k}$
$P_k = -\frac{\partial F'}{\partial Q_k}$	$Q_k = \frac{\partial F''}{\partial P_k}$
$H + \frac{\partial F'}{\partial t} = K$	$H + \frac{\partial F''}{\partial t} = K$.

Show that F' and F'' satisfy the following Hamilton-Jacobi equations

$$H\left(-\frac{\partial F'}{\partial p_1} \cdots p_1 \cdots\right) + \frac{\partial F'}{\partial t} = 0$$

$$H\left(-\frac{\partial F''}{\partial p_1} \cdots p_1 \cdots\right) + \frac{\partial F''}{\partial t} = 0 \quad .$$

These are less useful than (6.4a) and (6.4b) since the Hamiltonian usually depends on q in a more complicated way than it does on p.

14. Show that

$$[q(t),\ q(0)] = -\frac{\partial q(t)}{\partial p(0)}$$

$$[q(t),\ p(0)] = \frac{\partial q(t)}{\partial q(0)}$$

and therefore

$$\big(q(t),\ q(0)\big) = -\,i\hbar\,\frac{\partial q(t)}{\partial p(0)}$$

$$\big(q(t),\ p(0)\big) = i\hbar\,\frac{\partial q(t)}{\partial q(0)} \quad .$$

Check for an oscillator.

15. For the harmonic oscillator show that

$$\big(q(t_2),\ q(t_1)\big) = \frac{i\hbar}{\left(\dfrac{\partial^2 S}{\partial q(t_2)\ \partial q(t_1)}\right)}$$

and also that

$$\langle q_2 t_2 | q_1 t_1 \rangle \sim \left(\frac{\partial^2 S}{\partial q_2 \, \partial q_1} \right)^{1/2} e^{iS/\hbar} \quad .$$

16. Prove (9.33) and (9.34).

17. Show that

$$\langle q_2 t_2 | q_1 t_1 \rangle = \int \langle q_2 t_2 | qt \rangle \, dq \, \langle qt | q_1 t_1 \rangle$$

if

$$\langle qt | q_1 t_1 \rangle =$$

$$= \left(\frac{m\omega}{2\pi \, i\hbar \, \sin \omega T} \right)^{1/2} e^{\frac{im\omega}{2\hbar \sin \omega T} [(q^2 + q_1{}^2) \cos \omega T - 2q \, q_i]}$$

where $T = t - t_1$, and thereby establish (9.30).

18. Consider the following canonical transformation:

$$\begin{aligned} \bar{q} &= aq + bp \\ \bar{p} &= cq + dp \end{aligned} \qquad \begin{vmatrix} a & b \\ c & d \end{vmatrix} = 1 \quad .$$

Find the eigenfunction of $\bar{q}$ as follows:

$$(aq + bp) | \bar{q}' \rangle = \bar{q}' | \bar{q}' \rangle$$

$$(aq + bp) \int | q' \rangle \, dq' \, \langle q' | \bar{q}' \rangle = \bar{q}' \; | \bar{q}' \rangle$$

or

$$\int |q'\rangle\, dq' \left(aq' + b\,\frac{\hbar}{i}\frac{d}{dq'}\right) \langle q'|\bar{q}'\rangle = \int |q'\rangle\, dq'\bar{q}'\, \langle q'|\bar{q}'\rangle$$

and therefore

$$\left(aq' + b\,\frac{\hbar}{i}\frac{d}{dq'}\right) \langle q'|\bar{q}'\rangle = \bar{q}'\, \langle q'|\bar{q}'\rangle \quad .$$

By solving this equation find the following representation of the original contact transformation:

$$\langle q'|\bar{q}'\rangle = A\, e^{i(S/\hbar)}$$

where

$$S = \frac{1}{b}\,[\bar{q}'q' - \frac{a}{2}\,(q')^2] \quad .$$

19. The canonical transformation that solves the oscillator problem is

$$q = q_o \cos \omega t + \frac{p_o}{m\omega} \sin \omega t$$

$$\frac{p}{m\omega} = - q_o \sin \omega t + \frac{p_o}{m\omega} \cos \omega t \quad .$$

By using the result of the preceding problem show that the transformation function $\langle q'|q_o'\rangle$ that solves the oscillator problem is

$$\langle q'|q_o'\rangle = A\, e^{iS/\hbar}$$

where

$$S = \left(\frac{\omega}{2 \sin \omega t}\right) [(q'^2 + q_o'^2) \cos \omega t - 2q'q_o']$$

as in problem 17.

M. Moshinsky and C. Quesne, J. Math. Phys. 12, 1772 (1971).

20. Define the determinant

$$D = \left| \frac{\partial^2 S}{\partial q^m(t_2)\, \partial q^n(t_1)} \right| \quad .$$

Show that D satisfies the equation of continuity

$$\frac{\partial D}{\partial t_2} + \frac{\partial}{\partial q_2{}^m} (\dot{q}_2{}^m D) = 0 \quad ,$$

J. H. van Vleck, Proc. Natl. Acad. Sci. 14, 178 (1928).

21. Show that the one dimensional classical approximation to (7.3) is

$$\sim \frac{1}{\sqrt{p}}\, e^{-\frac{i}{\hbar} Et}\, e^{\pm \frac{i}{\hbar} \int p dx} \quad .$$

If the motion takes place in the potential illustrated in the figure, prove the quantization condition of the old quantum theory

$$\oint p(E, x)dx = \left(n + \frac{1}{2}\right) h$$

where the integral is taken over the whole period of the quasi-classical motion (from a to b and back).

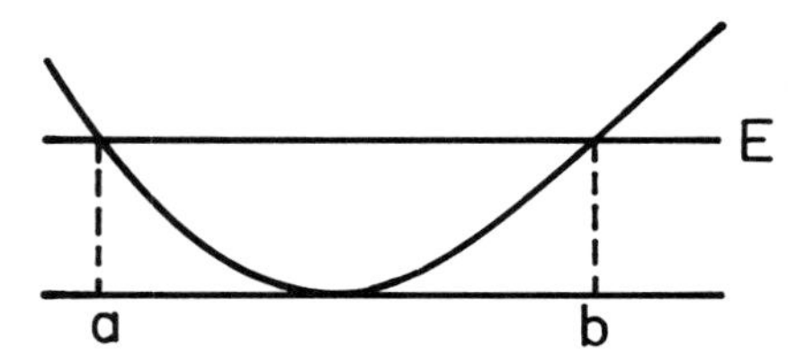

See Landau and Lifshitz, Quantum Mechanics, page 160. For the W.B.K. connection formulas see also E. C. Kemble, Fundamental Principles of Quantum Mechanics, page 97.

CHAPTER 4

DYNAMICS OF RIGID BODIES

4.1 THE TRACTABLE PROBLEMS OF DYNAMICS

The general methods so far described provide different viewpoints for discussing any particular dynamical system. Unless the given system exhibits some kind of symmetry, however, there is little one can do to solve the equations of motion (except numerically) or even to obtain more physical insight than can be gained from these general methods.

The exactly soluble problems of dynamics, e.g., the Kepler problem, are characterized by an invariance group that contains enough symmetry operations to yield a complete set of integrals of the motion. Problems that are not exactly soluble, but still tractable, are characterized by a Hamiltonian that may be written as a sum of two parts

$$H = \overset{\circ}{H} + V(\lambda)$$

where $\overset{\circ}{H}$ is invariant under a symmetry group and leads to an exactly soluble problem, while $V(\lambda)$ may be expanded in powers of the parameter and may be regarded as small. The small term $V(\lambda)$ then "breaks" the symmetry by a small amount. It is then possible to devise perturbation methods, based on an expansion in λ, for solving the problem associated with the complete Hamiltonian, H. For example, one may describe the solar system by first solving the Kepler problem for each planet separately and then calculating the perturbations $V(\lambda)$ of the planets on each other. Complex atoms may be analyzed in the same general way. The physics of any given problem then consists in an exact characterization of the perturbation $V(\lambda)$ and of the detailed way in which it breaks the original symmetry of $\overset{\circ}{H}$.

The central problem of this book is the analysis of the N-body non-relativistic system. In its general form this is not at all a tractable problem and one is able to analyze only restricted formulations. The simplest such restricted problem is in fact an idealization obtained by assuming that the N particles are frozen into a rigid solid with no vibrational motion. By studying such a model one can describe, for example, the precession and nutation of a planet and, at the other extreme, the rotational spectrum of a molecule. The next simple model is obtained by assuming a potential that permits the particles to vibrate about an equilibrium configuration. One obtains in this way the classical theory of vibrating systems and, in the quantum domain, the vibrational spectrum of a molecule. Finally, we shall discuss the problem that underlies our understanding of the planetary structure of the solar system as well as the electronic structure of atoms. In this chapter, however, we limit our discussion to the motion of rigid bodies.

4.2 GROUP OF RIGID BODY

The only changes possible for a free rigid body result from translations, rotations, and boosts from one state of uniform motion to another. Rotations and translations together generate the invariance group of Euclidian geometry; the generators of uniform motion are needed to complete the Galilean group.

The equations of motion of a free rigid body simply state that the generators of the Galilean group are time independent.

$$\frac{d}{dt} \underset{\sim}{P} = 0 \tag{2.1}$$

$$\frac{d}{dt} \underset{\sim}{L} = 0 \tag{2.2}$$

$$\frac{d}{dt}\mathbf{G} = 0 \quad . \tag{2.3}$$

The first two vector equations express the conservation of linear momentum and angular momentum respectively. The third equation gives

$$\mathbf{G} = \sum m_\alpha \mathbf{x}_\alpha - t\mathbf{P} = \text{const} \quad . \tag{2.4}$$

Define the center of mass ($\mathbf{X}$)

$$M\mathbf{X} = \sum m_\alpha \mathbf{x}_\alpha \quad . \tag{2.5}$$

Then

$$\mathbf{X} = (t/M)\mathbf{P} + \mathbf{X}_o \quad . \tag{2.6}$$

Therefore, the condition (2.3) states that the center of mass moves uniformly, when (2.1) holds.

One may choose a frame such that

$$\mathbf{P} = 0 \quad . \tag{2.7}$$

Then

$$\mathbf{X} = \mathbf{X}_o \quad . \tag{2.8}$$

It is convenient to calculate the angular momentum about this origin which is the center of mass.

If this co-moving frame, attached to the uniformly

moving center of mass, is also not rotating, then it is an inertial frame. On the other hand, if the coordinate system is attached to the rotating rigid body, it is in general also rotating and therefore is not an inertial frame and the dynamical equations no longer hold without the addition of fictitious forces.

If we were completely systematic at this point, we should next discuss the complete Galilean group. For our purposes, however, the analysis of the rotation group will suffice.

4.3 THE ROTATION GROUP

The product of two rotations is again a rotation:

$$R_1R_2 = R_3 \quad . \tag{3.1}$$

This is the group property. A group in the mathematical sense also has the following properties. There is an identity element, I. Each element has an inverse: $AA^{-1} = A^{-1}A = I$. The associative law holds: $A(BC) = (AB)C$. The set of all rotations clearly has these properties and is therefore a mathematical group.[1]

a) Matrix Representations

One obtains a representation of this group by associating

a matrix D(R) with each rotation (R) such that

$$D(R_1)\ D(R_2) = D(R_1 R_2) \quad . \tag{3.2}$$

The dimensionality of the representation is the dimensionality of the matrix D(R).

A one dimensional representation is obtained by simply putting D(R) = 1 for all R.

A three dimensional representation of the rotation group may be read from the vector transformation law:

$$X_k' = \sum R_{k\ell}\ X_\ell \quad . \tag{3.3}$$

Then the matrix $||R_{k\ell}||$ is associated with the rotation R which carries $\underset{\sim}{X}$ into $\underset{\sim}{X}'$.

If D(R) is a representation, then $U^{-1}D(R)U$ is also a representation. For

$$\begin{aligned} U^{-1}\ D(R_1)\ U \cdot U^{-1}\ D(R_2)\ U &= U^{-1}\ D(R_1)\ D(R_2)\ U \\ &= U^{-1}\ D(R_1 R_2)\ U \quad . \end{aligned}$$

Representations connected by a similarity transformation are called equivalent.

b) <u>Irreducible Representation</u>

Let D(R) be an arbitrary representation of the rotation

group. Suppose all D(R) can be simultaneously reduced to a step structure by the same similarity transformation U as follows:

$$U^{-1} D(R) U = \qquad (3.4)$$

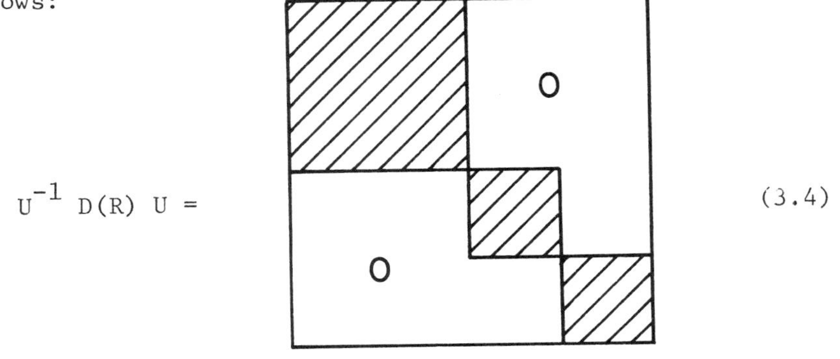

for all R. Then the representation D(R) is called reducible. Otherwise D(R) is called irreducible.

Every reducible representation may be transformed by a similarity transformation into a step structure in which each step is an irreducible representation. We denote the irreducible representations of the rotation group by $D^j(R)$ where $j = 0, 1/2, 1, \ldots$. The dimensionality of $D^j(R)$ is $2j + 1$. It may be shown that there is one irreducible representation for every $j = 0, 1/2, 1, \ldots$, and that all irreducible representations are included in this series.

c) Two-Dimensional Irreducible Representation

This is the simplest nontrivial irreducible representation. It is denoted by $D^{1/2}(R)$ and is called the spin representation.

Define

$$X = \begin{bmatrix} z & x-iy \\ x+iy & -z \end{bmatrix}$$

$$= x\sigma_1 + y\sigma_2 + z\sigma_3 = \underset{\sim}{x}\underset{\sim}{\sigma} \qquad (3.5)$$

where $\underset{\sim}{\sigma}$ is the set of the three Pauli matrices. Then the trace of X vanishes:

$$\text{Tr } X = 0 \quad . \qquad (3.6)$$

X is hermitian:

$$X^+ = X \qquad (3.7)$$

and its determinant is

$$\det X = -(x^2 + y^2 + z^2) \quad . \qquad (3.8)$$

Let U be a unitary 2×2 complex matrix such that

$$\det U = 1 \quad . \qquad (3.9)$$

U then depends on only three parameters.

Consider

$$X' = U X U^{-1} \quad . \qquad (3.10)$$

Then

$$\text{Tr } X' = \text{Tr } X = 0 \tag{3.11}$$

$$(X')^+ = X' \tag{3.12}$$

and

$$\det X' = \det X \quad . \tag{3.13}$$

It follows from (3.11) and (3.12) that X' may be written as follows

$$X' = \begin{pmatrix} z' & x'-iy' \\ x'+iy' & -z' \end{pmatrix} \tag{3.14}$$

where x', y', z' are real numbers. Then (3.13) and (3.8) imply

$$x^2 + y^2 + z^2 = x'^2 + y'^2 + z'^2 \quad . \tag{3.15}$$

Since (3.10) is a linear relation between $\underset{\sim}{x}$ and $\underset{\sim}{x}'$, it follows from (3.15) that this matrix transformation $X \to X'$ induces the rotation

$$x'_k = \sum R_{k\ell}\, x_\ell \quad . \tag{3.16}$$

Equations (3.5) and (3.10) may now be combined

$$\sum x'_k\, \sigma_k = \sum x_m\, U\, \sigma_m\, U^{-1} \quad .$$

By (3.16)

$$\sum R_{k\ell}\, x_\ell\, \sigma_k = \sum x_\ell\, U\, \sigma_\ell\, U^{-1}$$

and therefore

$$\sum \sigma_k R_{k\ell} = U \sigma_\ell U^{-1} \quad . \tag{3.17}$$

To verify that U is a representation one may show that

$$[U(1)U(2)] \sigma_\ell [U(1)U(2)]^{-1} = \sum \sigma_k [R(1)R(2)]_{k\ell} \quad . \tag{3.18}$$

The relation just written may be checked as follows:

$$\begin{aligned}
[U(1)U(2)] \sigma_\ell [U(1)U(2)]^{-1} &= U(1) [U(2) \sigma_\ell U(2)^{-1}] U(1)^{-1} \\
&= U(1) [\sum \sigma_k R_{k\ell}(2)] U(1)^{-1} \\
&= \sum [U(1) \sigma_k U(1)^{-1}] R_{k\ell}(2) \\
&= \sum_m \sigma_m [R(1)R(2)]_{m\ell} \quad .
\end{aligned}$$

Therefore U(R) is a representation of the rotation group. It is unitary by hypothesis, and it is also irreducible.[2]

The explicit relation between U and R may be found from (3.17) since

$$\mathrm{Tr}\, \sigma_m \sigma_n = \delta_{mn} \tag{3.19}$$

where Tr stands for $\frac{1}{2}$ x trace.

Then by (3.17) and (3.19)

$$R_{m\ell} = \mathrm{Tr}\, \sigma_m U \sigma_\ell U^{-1} \quad . \tag{3.20}$$

Notice that

$$R_{\ell m} = \mathrm{Tr}\ \sigma_\ell\ U\ \sigma_m\ U^{-1}$$

$$= \mathrm{Tr}\ \sigma_m\ U^{-1}\ \sigma_\ell\ U$$

by cyclical property of trace. Therefore

$$R_{\ell m}(U) = R_{m\ell}(U^{-1})$$

or

$$R^T = R^{-1} \tag{3.21}$$

where R^T means the transposed matrix.

Since U is unitary, one may write

$$U = e^{(i/2)\underset{\sim}{\sigma}\underset{\sim}{W}} \tag{3.22}$$

where $\underset{\sim}{W}$ is a real vector. How is $\underset{\sim}{W}$ related to the rotation which U represents? By (3.20) and (3.22)

$$R_{km} = \mathrm{Tr}\ \sigma_k\ e^{(i/2)\underset{\sim}{\sigma}\underset{\sim}{W}}\ \sigma_m\ e^{-(i/2)\underset{\sim}{\sigma}\underset{\sim}{W}} \quad . \tag{3.23}$$

Choose z-axis along $\underset{\sim}{W}$. Then

$$R_{km} = \mathrm{Tr}\ \sigma_k\ e^{(i/2)\sigma_3 W}\ \sigma_m\ e^{-(i/2)\sigma_3 W} \quad .$$

Then

$$R_{k3} = \mathrm{Tr}\ \sigma_k\ \sigma_3 = \delta_{k3} \quad . \tag{3.24}$$

If $k \neq 3$ and $m \neq 3$,

$$R_{km} = \mathrm{Tr}\, \sigma_k \sigma_m e^{-i\sigma_3 W}$$

$$= \mathrm{Tr}\, \sigma_k \sigma_m (\cos W - i \sigma_3 \sin W)$$

$$= \cos W\, \delta_{km} - i \sin W\, \mathrm{Tr}\, \sigma_k \sigma_m \sigma_3 \quad .$$

But

$$\mathrm{Tr}\, \sigma_k \sigma_m \sigma_\ell = i\, \varepsilon_{km\ell} \quad .$$

Therefore

$$R_{km} = (\cos W)\, \delta_{km} + (\sin W)\, \varepsilon_{km3} , \quad k,m \neq 3 \quad . \tag{3.25}$$

The rotation matrix is then

$$R_{k\ell} = \begin{pmatrix} \cos W & \sin W & 0 \\ -\sin W & \cos W & 0 \\ 0 & 0 & 1 \end{pmatrix} \quad . \tag{3.26}$$

Therefore R is a rotation about the direction of $\underset{\sim}{W}$ by amount W, and its two dimensional representation is

$$U = e^{(i/2)\underset{\sim}{\sigma}\underset{\sim}{W}} = \cos \frac{W}{2} + i\underset{\sim}{\sigma}\underset{\sim}{\hat{W}} \sin \frac{W}{2} \tag{3.27}$$

where $\hat{W}$ is the unit vector along $\underset{\sim}{W}$. Notice that (3.26) describes either a clockwise rotation of $\underset{\sim}{x}$ or a counterclockwise rotation of the frame about the axis of rotation.

d) Conjugate Rotations

Let R_1 and R_2 be two rotations. Consider

$$R_3 = R_2 R_1 R_2^{-1} \quad . \tag{3.28}$$

R_1 and R_3 are said to be conjugate. By the group property R_3 must also be a rotation. Let U_1, U_2, and U_3 be the corresponding two dimensional representations. Then

$$U_3 = U_2 U_1 U_2^{-1} = U_2 \left(e^{(i/2)\underset{\sim\sim}{\sigma W_1}}\right) U_2^{-1}$$
$$= \exp\left(\frac{i}{2} U_2 (\underset{\sim\sim}{\sigma W_1}) U_2^{-1}\right) \quad .$$

Therefore[3]

$$\underset{\sim\sim}{\sigma W_3} = U_2 (\underset{\sim\sim}{\sigma W_1}) U_2^{-1} \quad . \tag{3.29}$$

According to (3.10) and (3.16) the vector $\underset{\sim}{W}_3$ is obtained from $\underset{\sim}{W}_1$ by the rotation R_2. Therefore R_3 is a rotation of the same magnitude as R_1 with an axis obtained by rotating the axis of R_1 by R_2. Therefore conjugate rotations are of the same magnitude but have different axes.

If U_2 is near the identity, then $\underset{\sim}{W}_2$ is infinitesimal. Let the associated matrix $\underset{\sim\sim}{\sigma W_2}$ be denoted by W_2. Then

$$W_3 = W_1 + \frac{i}{2} (W_2, W_1)$$

and

$$\delta W = \frac{1}{2} i \; (W_2, \; W) \qquad .$$

This relation may be expressed in vector form by anticipating (4.9) of this chapter

$$\delta \underset{\sim}{W} = - \underset{\sim}{W}_2 \times \underset{\sim}{W} \qquad . \tag{3.29a}$$

e) Euler's Angles

An arbitrary rotation $\underset{\sim}{W}$ may be decomposed into a product of three rotations performed in the following order:

1. γ about z-axis,
2. β about y-axis,
3. α about z-axis.

Denote the product of the corresponding two dimensional matrices by $D^{1/2}(\alpha\beta\gamma)$. Then

$$D^{1/2}(\alpha\beta\gamma) = e^{(i/2)\sigma_3\alpha} \; e^{(i/2)\sigma_2\beta} \; e^{(i/2)\sigma_3\gamma} \tag{3.30}$$

$$= \begin{pmatrix} e^{(i/2)\alpha} & 0 \\ 0 & e^{-(i/2)\alpha} \end{pmatrix} \begin{pmatrix} \cos\frac{\beta}{2} & \sin\frac{\beta}{2} \\ -\sin\frac{\beta}{2} & \cos\frac{\beta}{2} \end{pmatrix} \begin{pmatrix} e^{(i/2)\gamma} & 0 \\ 0 & e^{(-i/2)\gamma} \end{pmatrix}$$

$$= \begin{bmatrix} e^{(i/2)(\alpha+\gamma)} \cos\frac{\beta}{2} & e^{(i/2)(\alpha-\gamma)} \sin\frac{\beta}{2} \\ -e^{-(i/2)(\alpha-\gamma)} \sin\frac{\beta}{2} & e^{-(i/2)(\alpha+\gamma)} \cos\frac{\beta}{2} \end{bmatrix} . \tag{3.31}$$

An arbitrary orientation of a rigid body may be described

with the aid of these angles. We imagine two coordinate frames which initially coincide and suppose that one of these frames, which is attached to the rigid body, is subjected to the three successive rotations described in (3.30) about the three <u>fixed</u> axes. One gets the same final position of the rigid body if the three successive rotations are made in <u>reverse</u> order about the <u>moving</u> axes. That is,

$$R_z(\alpha)\ R_y(\beta)\ R_z(\gamma) = R_{z'}(\gamma)\ R_{y'}(\beta)\ R_{z'}(\alpha) \quad . \qquad (3.32)$$

Here the prime refers to the body frame so that $R_y(\beta)$ and $R_{y'}(\beta)$ for example mean rotations of α about the y-axis of the fixed and body frames, respectively.[4] It is also of interest to compute $R_{m\ell}$ by (3.20) when U is expressed in Eulerian angles.

Let

$$U = D^{1/2}(\alpha,\beta,\gamma) = y_o + i\underset{\sim}{y}\underset{\sim}{\sigma} \quad . \qquad (3.33)$$

Since U is unimodular (det U = 1), the following relation must hold:

$$y^2 + y_o{}^2 = 1 \quad . \qquad (3.34)$$

Then one finds by (3.20)

$$R_{\ell m} = \delta_{\ell m}(1-2y^2) + 2y_\ell y_m + 2y_o y_s \varepsilon_{s\ell m} \quad . \qquad (3.35)$$

The connection between $(y_o, \underset{\sim}{y})$ and (α,β,γ) is by (3.31) and (3.33)

$$y_o = \cos\frac{1}{2}(\alpha+\gamma)\cos\frac{\beta}{2}$$

$$y_1 = \sin\frac{1}{2}(\alpha-\gamma)\sin\frac{\beta}{2}$$

$$y_2 = \cos\frac{1}{2}(\alpha-\gamma)\sin\frac{\beta}{2}$$

$$y_3 = \sin\frac{1}{2}(\alpha+\gamma)\cos\frac{\beta}{2} \quad . \qquad (3.36)$$

When (3.36) is substituted in (3.35) one finds the matrix elements of R. These are given in Table 3.1.

Table 3.1

$1 - 2y_2^2 - 2y_3^2$ $\cos\alpha\cos\beta\cos\gamma - \sin\alpha\sin\gamma$	$2y_1y_2 + 2y_0y_3$ $\sin\alpha\cos\gamma + \cos\alpha\cos\beta\sin\gamma$	$2y_1y_3 - 2y_0y_2$ $-\sin\beta\cos\alpha$
$2y_2y_1 - 2y_0y_3$ $-\cos\alpha\sin\gamma - \sin\alpha\cos\beta\cos\gamma$	$1 - 2y_1^2 - 2y_3^2$ $\cos\alpha\cos\gamma - \sin\alpha\cos\beta\sin\gamma$	$2y_2y_3 + 2y_0y_1$ $\sin\beta\sin\alpha$
$2y_3y_1 + 2y_0y_2$ $\cos\gamma\sin\beta$	$2y_3y_2 - 2y_0y_1$ $\sin\gamma\sin\beta$	$1 - 2y_1^2 - 2y_2^2$ $\cos\beta$

f) Generators of Rotation Group

Write the two dimensional representation in the following form

$$D^{1/2}(\alpha\beta\gamma) = e^{iG_3\alpha}\, e^{iG_2\beta}\, e^{iG_3\gamma} \tag{3.37}$$

where

$$\underset{\sim}{G} = \frac{1}{2}\underset{\sim}{\sigma} \quad . \tag{3.38}$$

One verifies that

$$(G_k, G_\ell) = i\, \varepsilon_{k\ell m}\, G_m \quad . \tag{3.39}$$

These are the commutation rules of the quantum mechanical angular momentum operator with $\hbar = 1$.

There is an infinite set of solutions of the commutation rules (3.39); denote a solution of dimensionality 2J+1 by $\underset{\sim}{G}^{(J)}$. From each solution $\underset{\sim}{G}^{(J)}$ one may construct

$$D^{(J)}(\alpha\beta\gamma) = e^{iG_3{}^{(J)}\alpha}\, e^{iG_2{}^{(J)}\beta}\, e^{iG_3{}^{(J)}\gamma} \quad . \tag{3.40}$$

It may be shown that all the irreducible unitary representations of the rotation group may be written in this way.[5] At the same time one sees how the irreducible representations of the rotation group $D^J(\alpha\beta\gamma)$ are related to the generators $\underset{\sim}{G}^J$.

Except for Planck's constant, $\underset{\sim}{G}$ is the same as the angular momentum operator $\underset{\sim}{J}$.

$$\underset{\sim}{J} = \hbar \underset{\sim}{G} \quad . \tag{3.41}$$

4.4 ROTATING FRAMES

Consider the rotation of a vector according to the equations

$$A(t) = U_t \; B \; U_t^{-1} \tag{4.1}$$

$$A_k(t) = \sum R_{k\ell}(t) \; B_\ell \tag{4.1a}$$

where $A = \underset{\sim}{A}\underset{\sim}{\sigma}$, $B = \underset{\sim}{B}\underset{\sim}{\sigma}$, U_t depends on the time, and $A(0) = B$. We may interpret these equations by supposing that a vector with components B_k is rigidly attached to a rotating frame that coincides at $t = 0$ with a certain fixed frame. Then $A_k(t)$ are the components of this same vector in the fixed frame at any later time. If we put $U = D^{1/2}(\alpha,\beta,\gamma)$ where (α,β,γ) is the sequence of rotations specified by (3.30), then the positive exponents in (3.30) would imply a sequence of clockwise rotations of the attached frame. Since we wish to conform to the usual conventions about anticlockwise Eulerian rotations, we shall put instead

$$U = D^{1/2}(-\alpha,-\beta,-\gamma) \quad . \tag{4.2}$$

Differentiate (4.1). Then

$$\frac{dA}{dt} = \frac{dU}{dt} B U^{-1} + U B \frac{dU^{-1}}{dt} + U \frac{dB}{dt} U^{-1}$$

$$= \left(\frac{dU}{dt} U^{-1}\right) A + A \left(U \frac{dU^{-1}}{dt}\right) + U \frac{dB}{dt} U^{-1} \quad . \qquad (4.3)$$

Define

$$\Omega = i \frac{dU}{dt} U^{-1} \quad . \qquad (4.4)$$

Then by (4.3)

$$\frac{dA}{dt} = U \frac{dB}{dt} U^{-1} + i(A,\Omega) \quad . \qquad (4.5)$$

The work leading to (4.5) is formally the same as that used to discuss the relation of the Schrödinger to the Heisenberg picture.

Equation (4.5) may be written in vector form if we introduce the vector with components $\overset{\circ}{\omega}_k$ in the fixed frame such that

$$\Omega = \frac{1}{2} \overset{\circ}{\underset{\sim}{\omega}} \underset{\sim}{\sigma} \qquad (4.6)$$

and make use of the identities connecting any two vectors, say $\underset{\sim}{U}$ and $\underset{\sim}{V}$, and their matrices $U = \underset{\sim}{U}\underset{\sim}{\sigma}$ and $V = \underset{\sim}{V}\underset{\sim}{\sigma}$, namely[6]:

$$UV = \underset{\sim}{U}\underset{\sim}{V} + i\, \underset{\sim}{\sigma} \cdot \underset{\sim}{U} \times \underset{\sim}{V} \qquad (4.7)$$

$$\frac{1}{2}(UV + VU) = \underset{\sim\sim}{UV} \tag{4.8}$$

$$\frac{1}{2}(UV - VU) = i\,\underset{\sim}{\sigma} \cdot \underset{\sim}{U} \times \underset{\sim}{V} \quad . \tag{4.9}$$

Then (4.5) becomes

$$\underset{\sim\sim}{\dot{A}\sigma} = (U\underset{\sim}{\sigma}U^{-1})\,\underset{\sim}{\dot{B}} + \underset{\sim}{\sigma}(\underset{\sim}{\overset{\circ}{\omega}} \times \underset{\sim}{A}) \quad .$$

With the aid of (3.17) one finds

$$\dot{A}_m = (\underset{\sim}{\overset{\circ}{\omega}} \times \underset{\sim}{A})_m + \sum_k R_{mk}\,\dot{B}_k \quad . \tag{4.10a}$$

Here $\dot{B}_k$ is the rate of change with respect to the body frame. The total rate of change with respect to the fixed (or inertial) frame is the sum of the first term, which is caused by the rotation of the body frame, and this second term, which vanishes if the vector is fixed in the body frame.

If the body and fixed frame instantaneously coincide, one may write

$$\frac{d}{dt}\underset{\sim}{A} = \frac{D}{Dt}\underset{\sim}{A} + \underset{\sim}{\overset{\circ}{\omega}} \times \underset{\sim}{A} \tag{4.11}$$

where D/Dt is the derivative with respect to the moving axes.

The preceding familiar equation is obtained in a simpler way by vector analysis,[7] since in this case it is easier to work with the three dimensional representations from the beginning. [Compare (4.11) with (3.29a).] Finally the right side of (4.10a) may be expressed entirely in the rotating frame as follows:

$$\sum_m R^{-1}_{nm} \dot{A}_m = \sum_m R^{-1}_{nm} (\overset{\circ}{\underset{\sim}{\omega}} \times \underset{\sim}{A})_m + \dot{B}_n$$

$$= (\underset{\sim}{\bar{\omega}} \times \underset{\sim}{B})_n + \dot{B}_n \tag{4.10b}$$

where $\underset{\sim}{\bar{\omega}}$ is also referred to the body frame.

Let us next compute the angular velocity in terms of the Eulerian angles by combining (4.4) and (4.6)

$$\Omega = -i\, U \frac{dU^{-1}}{dt} = \frac{1}{2} \overset{\circ}{\underset{\sim}{\omega}} \underset{\sim}{\sigma} \quad . \tag{4.12}$$

Therefore

$$\overset{\circ}{\omega}_k = -2i \,\mathrm{Tr}\, U \frac{dU^{-1}}{dt} \sigma_k \quad . \tag{4.13}$$

For U one now has, by (4.2):

$$U = e^{-(i/2)\sigma_3\alpha}\, e^{-(i/2)\sigma_2\beta}\, e^{-(i/2)\sigma_3\gamma} \quad . \tag{4.14}$$

A direct calculation, making use of (4.13) and (4.14), gives in the inertial frame:

$$\overset{\circ}{\omega}_1 = \dot{\gamma} \sin\beta \cos\alpha - \dot{\beta} \sin\alpha \tag{4.15a}$$

$$\mathring{\omega}_2 = \dot{\gamma} \sin\beta \sin\alpha + \dot{\beta} \cos\alpha \tag{4.15b}$$

$$\mathring{\omega}_3 = \dot{\gamma} \cos\beta + \dot{\alpha} \quad . \tag{4.15c}$$

The components $\bar{\omega}_k$ of $\underset{\sim}{\omega}$ in the body frame are related to $\mathring{\omega}_k$ by

$$\mathring{\omega}_k = \sum R_{k\ell}(-\alpha,-\beta,-\gamma)\, \bar{\omega}_\ell \tag{4.16a}$$

according to (4.1a), since $A_k(t)$ and B_k describe the same vector in the two frames if B_k is time-independent.

Then

$$\bar{\omega}_\ell = \sum R^{-1}_{\ell k}\, \mathring{\omega}_k = \sum_k \mathring{\omega}_k\, R_{k\ell} \tag{4.16}$$

$$= -2i \sum_k \mathrm{Tr}\, U \frac{dU^{-1}}{dt} \sigma_k R_{k\ell} \quad \text{by} \quad (4.13)$$

$$= -2i\ \mathrm{Tr}\, U \frac{dU^{-1}}{dt} U \sigma_\ell U^{-1} \quad \text{by} \quad (3.17)$$

$$= 2i\ \mathrm{Tr}\, U^{-1} \frac{dU}{dt} \sigma_\ell \quad . \tag{4.17}$$

By comparing (4.17) with (4.13) one sees that

$$\bar{\omega}_\ell(\alpha,\beta,\gamma) = -\,\mathring{\omega}_\ell(-\gamma,-\beta,-\alpha) \quad . \tag{4.18}$$

Therefore in the body frame[8]

$$\bar{\omega}_1 = -\,\dot{\alpha} \sin\beta \cos\gamma + \dot{\beta} \sin\gamma$$

$$\bar{\omega}_2 = \dot{\alpha} \sin\beta \sin\gamma + \dot{\beta} \cos\gamma$$

$$\bar{\omega}_3 = \dot{\alpha} \cos\beta + \dot{\gamma} \quad . \tag{4.19}$$

The components (4.19) in the body frame may also be computed directly from (4.16) and Table 3.1 for $R_{k\ell}$. They may also be obtained by geometric inspection of the components of the vector angular velocity in the body frame.

4.5 DYNAMICS OF RIGID BODY

a) Angular Momentum

Consider the rotation of a rigid body about an internal point. Let $\underset{\sim}{r}$ be the distance of any particle from this point. Then $\underset{\sim}{r}$ is constant with respect to any body fixed frame. With respect to an inertial frame

$$\frac{d}{dt} \underset{\sim}{r} = \underset{\sim}{\omega} \times \underset{\sim}{r} \tag{5.1}$$

where $\underset{\sim}{\omega}$ is written for $\overset{\circ}{\underset{\sim}{\omega}}$. As a matrix equation (5.1) reads

$$i \frac{dR}{dt} = (\Omega, R) \tag{5.2}$$

where $R = \underset{\sim}{r}\underset{\sim}{\sigma}$ and $\Omega = \frac{1}{2} \underset{\sim}{\omega}\underset{\sim}{\sigma}$. Then

$$iP = m(\Omega, R) \tag{5.3}$$

where P is the momentum matrix:

$$P = m\left(\frac{d}{dt}\underset{\sim}{r}\right)\underset{\sim}{\sigma} \quad . \tag{5.4}$$

Now the angular momentum of a particle is

$$\underset{\sim}{\ell} = \underset{\sim}{r} \times \underset{\sim}{p} \tag{5.5}$$

and therefore

$$iL = \frac{1}{2}(R, P) \quad . \tag{5.6}$$

Consequently by (5.3) and the preceding equation

$$L = \frac{m}{2}\big((\Omega, R), R\big) \quad . \tag{5.7}$$

The double commutator corresponds to the double vector product and has a similar expansion[6]

$$\begin{aligned} L &= \frac{m}{2}(\Omega R^2 - 2R\Omega R + R^2\Omega) \\ &= m(r^2\Omega - R\Omega R) \quad . \end{aligned} \tag{5.8}$$

But

$$\Omega R = \{\Omega, R\} - R\Omega = \underset{\sim}{\omega}\underset{\sim}{r} - R\Omega \tag{5.9}$$

where

$$\{\Omega, R\} = R\Omega + \Omega R = \underset{\sim}{\omega}\underset{\sim}{r} \quad . \tag{5.10}$$

Therefore by (5.8) and (5.9)

$$L = m\left(2r^2\Omega - (\underset{\sim}{r}\underset{\sim}{\omega})R\right) \tag{5.11}$$

and

$$\ell_k = \sum_s m(r^2\delta_k{}^s - r^s r_k)\ \omega_s \quad . \tag{5.12}$$

This relation holds for each particle in the rigid body. The total angular momentum is then

$$L_k = \sum_s I_k{}^s\ \omega_s \tag{5.13}$$

where

$$I_k{}^s = \sum_\alpha m_\alpha(r_\alpha{}^2\delta_k{}^s - r_{\alpha k}r_\alpha{}^s) \tag{5.14}$$

and the sum is over all particles α in the rigid body. Then the total angular momentum is determined by the tensor I_{ks} and the angular velocity ω_s, which is the same for all particles in the rigid body.

I_{ks} is known as the inertia tensor. It defines the ellipsoid

$$I_{ks}\ x^k\ x^s = 1 \quad . \tag{5.15}$$

The axes of this ellipsoid are the principal axes of inertia, and the characteristic values of the matrix I_{ks} are the principal moments of inertia.

b) Kinetic Energy

The kinetic energy of a single particle is

$$T = \frac{p^2}{2m} \quad . \tag{5.16}$$

Then

$$T = \frac{1}{2m} \text{Tr } P\, P = \frac{i}{2} \text{Tr } P(R, \Omega) \tag{5.17}$$

by (5.3). Then by the cyclical property of the trace, (5.17) may be rewritten

$$T = \frac{i}{2} \text{Tr } (P, R)\Omega = \text{Tr } L\Omega$$

$$= \frac{1}{2} \underset{\sim}{\ell} \underset{\sim}{\omega} \quad . \tag{5.18}$$

The preceding relation holds for every particle in the rigid body. One may now sum over all particles to obtain

$$T = \sum T_\alpha = \frac{1}{2} \left(\sum \underset{\sim}{\ell}_\alpha\right) \underset{\sim}{\omega}$$

or

$$T = \frac{1}{2} \underset{\sim}{L} \underset{\sim}{\omega} \quad . \tag{5.19}$$

Therefore the total kinetic energy is simply determined by the total angular momentum. Combining (5.19) and (5.13) one finds

$$T = \frac{1}{2} \sum I_{sk} \omega_s \omega_k \quad . \tag{5.20}$$

c) Eulerian Equations of Motion of Free Body

Since the linear momentum is conserved, the center of mass moves uniformly. Therefore a non-rotating inertial frame may be attached to the center of mass and in this frame the angular momentum is also constant

$$\frac{d}{dt} \underset{\sim}{L} = 0 \quad . \tag{5.21}$$

Since the inertia tensor is not constant with respect to this inertial frame, it is convenient to choose a rotating frame attached to the rigid body (body frame). The inertia tensor is then constant but the rotating frame is no longer inertial, and therefore it is necessary to introduce fictitious forces. However, this procedure turns out to be very convenient and the equations of motion become by (4.10b)

$$\frac{d\underset{\sim}{K}}{dt} + \underset{\sim}{\omega} \times \underset{\sim}{K} = 0 \tag{5.22}$$

where $d\underset{\sim}{K}/dt$, $\underset{\sim}{K}$, and $\underset{\sim}{\omega}$ are all referred to the rotating frame.

We are now using $\underset{\sim}{K}$ to explicitly indicate the angular momentum referred to the body frame. Notice that ω_k was denoted by $\bar{\omega}_k$ in (4.10b).

$$\frac{d}{dt} \underset{\sim}{K} = - \underset{\sim}{\omega} \times \underset{\sim}{K} \quad . \tag{5.25}$$

The right hand side of the equation describes a fictitious torque.

Now choose the body frame to lie along the principal axes of inertia. Then

$$I_{mn} = I_m \delta_{mn} \tag{5.24}$$

$$\dot{I}_{mn} = 0 \tag{5.25}$$

$$K_m = I_m \omega_m \tag{5.26}$$

$$T = \frac{1}{2} (I_1 \omega_1{}^2 + I_2 \omega_2{}^2 + I_3 \omega_3{}^2) \tag{5.27}$$

$$T = \frac{1}{2} \left(\frac{K_1{}^2}{I_1} + \frac{K_2{}^2}{I_2} + \frac{K_3{}^2}{I_3} \right) \quad . \tag{5.28}$$

The equations of motion become

$$\dot{K}_s + \varepsilon_{smn} \, \omega^m K^n = 0 \tag{5.29}$$

or

$$\dot{K}_1 + \omega_2 K_3 - \omega_3 K_2 = 0$$

$$\dot{K}_2 + \omega_3 K_1 - \omega_1 K_3 = 0$$

$$\dot{K}_3 + \omega_1 K_2 - \omega_2 K_2 = 0 \quad . \tag{5.29a}$$

Finally by (5.26)

$$\dot{\omega}_k = c_{kmn}\, \omega_m \omega_n \tag{5.30a}$$

where there is no sum on m and n, where kmn is the cyclic order, and where

$$c_{kmn} = \frac{I_m - I_n}{I_k} \quad . \tag{5.30b}$$

These are the Eulerian equations for a rigid body under no external forces. These equations hold only in a body frame oriented along the principal axes.

4.6 MOTION OF A RIGID BODY WHICH IS FREE

If the rigid body is free, then the angular momentum is constant in every inertial frame. However, the angular velocity is not constant. In general the plane of $\underset{\sim}{L}$ and $\underset{\sim}{\omega}$ will turn about $\underset{\sim}{L}$ and in addition, the angle between $\underset{\sim}{L}$ and $\underset{\sim}{\omega}$ will also change. These motions are called precession and nutation respectively. Finally the body is instantaneously rotating about $\underset{\sim}{\omega}$.

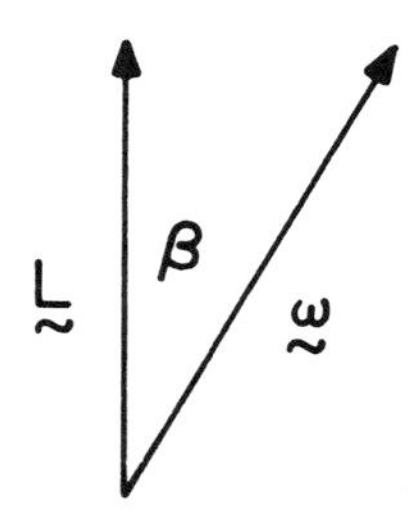

Fig. 6.1 Precession and Nutation. Rotation of plane of $\underset{\sim}{L}$ and $\underset{\sim}{\omega}$ about $\underset{\sim}{L}$ is precession. Variation of β is nutation.

The different possibilities may be classified according to the shape of the ellipsoid of inertia.

a) $I_1 = I_2 = I_3$ (Spherical Top)

According to (5.30)

$$\dot{\underset{\sim}{\omega}} = 0 \quad . \tag{6.1}$$

In fact

$$\underset{\sim}{\omega} = \frac{1}{I}\underset{\sim}{K} \quad . \tag{6.2}$$

Therefore $\underset{\sim}{\omega}$ is parallel to $\underset{\sim}{K}$ and constant if the body is spherical. If such a body is put into rotation about any axis there will be no tendency for this motion to change.

b) $I_1 = I_2 = I_\perp$ (Symmetric Top)

Then

$$\dot{\omega}_3 = 0$$

$$\dot{\omega}_2 = c_{231}\omega_3\omega_1$$

$$\dot{\omega}_1 = c_{123}\omega_2\omega_3 \tag{6.3}$$

and

$$\ddot{\omega}_2 = (c_{231}c_{123}\omega_3^{\,2})\omega_2$$

or

$$\ddot{\omega}_2 = -\,\lambda^2\,\omega_2 \tag{6.4}$$

where

$$\lambda = \left|\frac{I_3 - I_\perp}{I_\perp}\right| \omega_3 = \left|\frac{I_3 - I_\perp}{I_3 I_\perp}\right| K_3 \quad . \tag{6.5}$$

Therefore

$$\omega_3 = \text{constant}, \; \omega_2 = A \sin \lambda t, \; \omega_1 = A \cos \lambda t \quad . \tag{6.6}$$

The earth is an example of a free top for which ω_3 corresponds to a period of one day, $(I_\perp - I_3)/I_\perp \cong 1/300$ due to the flattening of earth at the poles, and therefore λ corresponds to a period of about 300 days.[9]

The constants, A and λ, are determined by the energy and angular momentum as follows:

$$2T = \frac{1}{I_3} K_3{}^2 + \frac{K_2{}^2 + K_1{}^2}{I_\perp} = \left(\frac{1}{I_3} - \frac{1}{I_\perp}\right) K_3{}^2 + \frac{K^2}{I_\perp}$$

and

$$\frac{|2T\, I_\perp - K^2|}{K_3 I_\perp} = \lambda \quad . \tag{6.7}$$

Finally

$$A^2 = \omega_2{}^2 + \omega_1{}^2 = (K_2{}^2 + K_1{}^2) \frac{1}{I_\perp{}^2} = \frac{K^2 - K_3{}^2}{I_\perp{}^2} \quad . \tag{6.8}$$

If such a body is put into rotation about any axis except I_3 the angular velocity $\underset{\sim}{\omega}$ will precess with the frequency λ about the principal axis.

To determine the motion of the top relative to an inertial frame, take the z-axis along the direction of the constant $\underset{\sim}{L}$. Then by Table 3.1 and (4.16a) the components of $\underset{\sim}{\omega}$ and $\underset{\sim}{K}$ in the body fixed frame are

$$K_1 = R^{-1}_{13}L = -L \sin\beta \cos\gamma = I_\perp \omega_1$$

$$K_2 = R^{-1}_{23}L = L \sin\beta \sin\gamma = I_\perp \omega_2$$

$$K_3 = R^{-1}_{33}L = L \cos\beta = I_3 \omega_3 \quad . \tag{6.9}$$

It follows from (6.6) that β is constant and $-\gamma = \lambda t$. It also follows from (6.6) and (4.19) that $\dot{\alpha}$ = constant.

c) $I_3 > I_2 > I_1$ <u>(Asymmetric Top)</u>

Now

$$2T - I_2\omega_2^{\,2} = I_1\omega_1^{\,2} + I_3\omega_3^{\,2} \tag{6.10}$$

$$K^2 - I_2^{\,2}\omega_2^{\,2} = I_1^{\,2}\omega_1^{\,2} + I_3^{\,2}\omega_3^{\,2} \quad . \tag{6.11}$$

These equations may be solved for $\omega_1^{\,2}$ and $\omega_3^{\,2}$ which are then linear functions of $\omega_2^{\,2}$

$$\omega_1^{\,2} = a + b\omega_2^{\,2} \tag{6.12}$$

$$\omega_3^{\,2} = c + d\omega_2^{\,2} \quad . \tag{6.13}$$

Then by (5.30)

$$\dot{\omega}_2 = \left(\frac{I_3 - I_1}{I_2}\right) [(a + b\omega_2^{\,2})(c + d\omega_2^{\,2})]^{1/2} \quad . \tag{6.14}$$

Equation (6.14) may be written in terms of the given energy and angular momentum as follows:

$$\dot{\omega}_2 = [(\beta_1\omega_2^{\,2} + \alpha_3)(\beta_3\omega_2^{\,2} - \alpha_1)]^{1/2} \tag{6.15}$$

where

$$\alpha_3 = \frac{L^2 - 2TI_3}{I_1 I_2} \qquad \beta_3 = \frac{I_2 - I_1}{I_3}$$

$$\alpha_1 = \frac{L^2 - 2TI_1}{I_2 I_3} \qquad \beta_1 = \frac{I_3 - I_2}{I_1} \quad .$$

Then the solution is the Jacobian elliptic function[8]

$$\omega_2 = (-\alpha_3/\beta_1)^{1/2} \, \mathrm{sn}(\alpha_1\beta_1)^{1/2} \, t \quad . \tag{6.16}$$

The complete solution is of the form

$$\omega_1 = A_1 \, \mathrm{cn} \, \tau \tag{6.17a}$$

$$\omega_2 = A_2 \, \mathrm{sn} \, \tau \tag{6.17b}$$

$$\omega_3 = A_3 \, \mathrm{dn} \, \tau \tag{6.17c}$$

where

$$\tau = (\alpha_1\beta_1)^{1/2} \, t \quad . \tag{6.18}$$

The period in t is

$$4K/(\alpha_1\beta_1)^{1/2} \quad . \tag{6.19}$$

To determine the motion of the top relative to an inertial frame, take the z-axis along the direction of the constant $\underset{\sim}{L}$ and use (6.9) again.

Then

$$\cos\beta = \frac{I_3\omega_3}{L} = \frac{I_3A_3}{L}\,\mathrm{dn}\,\tau \tag{6.20}$$

$$-\tan\gamma = \frac{I_2\omega_2}{I_1\omega_1} = \frac{I_2}{I_1}\,\frac{A_2}{A_1}\,\frac{\mathrm{sn}\,\tau}{\mathrm{cn}\,\tau} \quad . \tag{6.21}$$

These equations give β and γ as functions of the time. Finally by (4.19)

$$\dot{\alpha} = \frac{-\omega_1\cos\gamma + \omega_2\sin\gamma}{\sin\beta} \tag{6.22}$$

and therefore by (6.9)

$$\dot{\alpha} = \left(\frac{I_1\omega_1^{\,2} + I_2\omega_2^{\,2}}{I_1^{\,2}\omega_1^{\,2} + I_2^{\,2}\omega_2^{\,2}}\right) L \quad . \tag{6.23}$$

This equation may also be integrated[10] and one finds

$$\alpha(t) = \phi_1(t) + \phi_2(t) \tag{6.24}$$

where ϕ_1 and ϕ_2 may be expressed in terms of Jacobian theta functions. The period (T) of $\phi_1(t)$ is the same as the periods of $\beta(t)$ and $\gamma(t)$ but the period of $\phi_2(t)$ is not commensurable with T. Therefore the top does not at any time return to its initial position.

When $I_1 = I_2$, these results reduce to those obtained earlier for the symmetric top. For then sn → sin, cn → cos, and dn → 1. In addition one sees by (6.23) that $\dot{\alpha}$ = constant.

d) Stability

Suppose body is rotating about a principal axis. Then $\omega_2 = \omega_3 = 0$. Now assume a small perturbation so that $\omega_2 = \varepsilon$. Then

$$\ddot{\omega}_2 = \lambda\omega_2 \tag{6.25}$$

where

$$\lambda = c_{231}c_{312}\omega^2 \quad . \tag{6.25a}$$

The sign of λ depends on

$$c_{231}c_{312} = \frac{(I_3-I_1)(I_1-I_2)}{I_2 I_3} \quad . \tag{6.26}$$

If $I_3 > I_1 > I_2$, then $\lambda > 0$ and the motion is unstable. On the other hand, if I_1 is the greatest or least of the three principal moments, then $\lambda < 0$ and the motion is stable.

e) Poinsot Construction[11]

There is a simple qualitative way to describe the motion of an asymmetrical top. First return to the inertial frame.

Next define

$$I = \sum I_{mn} \hat{\omega}_m \hat{\omega}_n \tag{6.27}$$

where $\hat{\omega}$ is the unit vector along the axis of rotation. I is known as the moment of inertia about this axis. Then by (5.20)

$$T = \frac{1}{2} \sum I_{mn} \omega_m \omega_n = \frac{1}{2} I\omega^2 \quad . \tag{6.28}$$

Define

$$\rho_m = \frac{\hat{\omega}_m}{I^{1/2}} \quad . \tag{6.29}$$

Then

$$\sum I_{mn} \rho_m \rho_n = \frac{1}{I} \sum I_{mn} \hat{\omega}_m \hat{\omega}_n = 1 \quad . \tag{6.30}$$

Let

$$F(\rho) = \sum I_{mn} \rho_m \rho_n \quad . \tag{6.31}$$

Then (6.30) becomes

$$F(\rho) = 1 \quad . \tag{6.32}$$

The normal to this surface is parallel to $\underset{\sim}{\nabla} F$.

$$\frac{\partial F}{\partial \rho_m} = \sum 2\, I_{mn} \rho_n = \frac{2}{I^{1/2}} \sum I_{mn} \left(\frac{\omega_n}{\omega} \right) = \frac{2\, L_m}{I^{1/2}\, \omega} \quad . \tag{6.33}$$

Therefore the normal to the ellipsoid (6.32) is parallel to $\underset{\sim}{L}$. Since the angular momentum $\underset{\sim}{L}$ is fixed in space, it determines also an invariable plane perpendicular to it. Therefore, one may think of the ellipsoid (6.32) which is always tangent to this plane, as rolling on it. The rolling takes place without slipping since the point of contact lies on the axis of rotation $\hat{\omega}$ and is therefore instantaneously at rest.

Finally the height of the center of the ellipsoid above the invariable plane is also constant. For

$$\underset{\sim}{\rho}\hat{\underset{\sim}{L}} = \frac{\hat{\omega}}{I^{1/2}} \frac{\underset{\sim}{L}}{L} = \frac{\underset{\sim}{\omega}\underset{\sim}{L}}{I^{1/2}\omega L} = \frac{2T}{(I\omega^2)^{1/2}L}$$

$$= \frac{(2T)^{1/2}}{L} = \text{const.} \tag{6.34}$$

The point of contact traces out one curve on the inertial ellipsoid, called the polhode, and a second curve on the invariable plane, known as the herpolhode. The Poinsot construction is general and includes, for example, the case described analytically under (c).

4.7 RIGID BODY IN EXTERNAL FIELD

If the rigid body is not free, let the Hamiltonian be $\overset{\circ}{H} + V$, where V is a perturbation; then instead of (2.1) and

(2.2) one has

$$\frac{dP_k}{dt} = [P_k,H] = [P_k,V] = -\sum_\alpha \frac{\partial V}{\partial x_{k\alpha}} = \sum_\alpha F_{k\alpha} = F_k \qquad (7.1)$$

$$\begin{aligned} \frac{dL_{k\ell}}{dt} &= [L_{k\ell},H] = [L_{k\ell},V] \\ &= \sum_\alpha \left[x_{k\alpha}\left(-\frac{\partial V}{\partial x_{\ell\alpha}}\right) - x_{\ell\alpha}\left(-\frac{\partial V}{\partial x_{k\alpha}}\right) \right] \\ &= \sum_\alpha (x_{k\alpha}F_{\ell\alpha} - x_{\ell\alpha}F_{k\alpha}) = N_{k\ell} \end{aligned} \qquad (7.2)$$

where F_k is the force and $N_{k\ell}$ is the torque acting on the system. Here it has been assumed that the kinetic energy is still invariant under rotations and translations. The force and torque then appear only if the potential energy depends on position and orientation respectively. The generators of the rotation-translation group are then no longer constant and one may say that this symmetry is then broken.

One now has in an inertial frame

$$\frac{d\underset{\sim}{L}}{dt} = \underset{\sim}{N} \qquad (7.3)$$

instead of (5.22). Therefore in a body fixed frame, by (4.10b),

$$\frac{d\underset{\sim}{K}}{dt} = R^{-1}\,\underset{\sim}{N} - \underset{\sim}{\omega} \times \underset{\sim}{K} \qquad (7.4)$$

where $\underset{\sim}{N}$ is the real torque and $\underset{\sim}{\omega} \times \underset{\sim}{K}$ is again a fictitious torque associated with the rotating coordinate system.

The two terms on the right may be described as two kinds of symmetry breaking. $\underset{\sim}{N}$ represents a dynamical symmetry breaking while $\underset{\sim}{\omega} \times \underset{\sim}{K}$ is a symmetry breaking associated with a privileged coordinate system.

We next give a simple example of a rigid body in an external field.

a) Heavy Symmetric Top With One Point Fixed[12]

Consider a symmetric top spinning in a constant gravitational field and resting on the tip of its major axis. In the figure, the axis of the top is z'. Let the angle z0z' be β as shown; let the plane z0z' make the angle α with the plane x0z; and finally let γ be the amount by which the top is rotated about its own axis. Then (α,β,γ) are Eulerian angles.

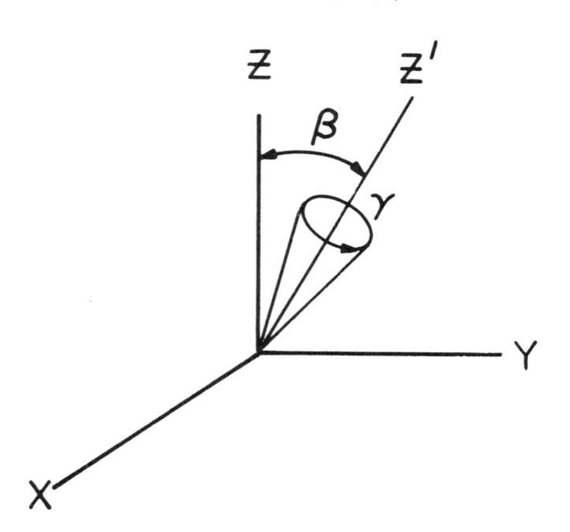

Figure 7.1. Top

The Lagrangian is

$$L = \frac{1}{2} I_{\perp}(\omega_1{}^2 + \omega_2{}^2) + \frac{1}{2} I_3\omega_3{}^2 - Mg\ell \cos\beta \tag{7.5}$$

where

$$\omega_1 = -\dot{\alpha} \sin\beta \cos\gamma + \dot{\beta} \sin\gamma$$

$$\omega_2 = \dot{\alpha} \sin\beta \sin\gamma + \dot{\beta} \cos\gamma$$

$$\omega_3 = \dot{\alpha} \cos\beta + \dot{\gamma} \tag{7.6}$$

are the components of $\underset{\sim}{\omega}$ in the body frame by (4.19). Then

$$L = \frac{1}{2} I_{\perp}(\dot{\alpha}^2 \sin^2\beta + \dot{\beta}^2) + \frac{1}{2} I_3(\dot{\alpha} \cos\beta + \dot{\gamma})^2 - Mg\ell \cos\beta \ . \tag{7.7}$$

Therefore

$$\frac{\partial L}{\partial\alpha} = \frac{\partial L}{\partial\gamma} = 0 \tag{7.8}$$

or α and γ are ignorable, and

$$\frac{d}{dt} p_\alpha = \frac{d}{dt} p_\gamma = 0 \tag{7.9}$$

where p_α and p_α are conjugate momenta. Therefore p_α and p_γ are constants of the motion. But

$$p_\alpha = \frac{\partial L}{\partial \dot{\alpha}} = I_\perp \dot{\alpha} \sin^2\beta + I_3(\dot{\alpha}\cos^2\beta + \dot{\gamma}\cos\beta)$$

$$= I_\perp b \tag{7.10}$$

$$p_\gamma = \frac{\partial L}{\partial \dot{\gamma}} = I_3(\dot{\alpha}\cos\beta + \dot{\gamma}) = I_\perp a \tag{7.11}$$

where a and b are constants. The last equation also implies that ω_3 is constant.

By combining (7.10) and (7.11) one gets

$$\dot{\alpha} = \frac{b - a\cos\beta}{\sin^2\beta} \quad . \tag{7.12}$$

There is also an energy integral:

$$\frac{1}{2} I_\perp(\omega_1{}^2 + \omega_2{}^2) + \frac{1}{2} I_3\omega_3{}^2 + Mg\ell\cos\beta = E \quad . \tag{7.13}$$

Since ω_3 is a constant, (7.13) may be written

$$\frac{1}{2} I_\perp(\omega_1{}^2 + \omega_2{}^2) + Mg\ell\cos\beta = E' \quad . \tag{7.14}$$

By (7.6) and (7.12) the energy integral may finally be expressed in the following equations:

$$(b - a\cos\beta)^2 + \dot{\beta}^2 \sin^2\beta = \frac{2}{I_\perp}(E' - Mg\ell\cos\beta)\sin^2\beta \quad . \tag{7.15}$$

Let

$$u = \cos\beta \quad .$$

Then (7.15) becomes

$$\dot{u}^2 = -(b - au)^2 + \frac{2}{I_\perp}(E' - Mg\ell\, u)(1 - u^2)$$

$$= f(u) \tag{7.16a}$$

where f(u) is a cubic in u or

$$\dot{u}^2 = A(u - u_1)(u - u_2)(u - u_3) \quad . \tag{7.16b}$$

There is one root $u_3 > 1$ (Fig. 7.2). The root does not lie in the physical region since u is the cosine of a real angle. Motion takes place in the zone between u_1 and u_2, in the sense that the axis of the top always pierces the sphere in this zone (Fig. 7.3). The variation of β in this zone is a nutational motion. In addition it follows from (7.12) that

$$\dot{\alpha} \sim \bar{u} - u \,, \qquad \text{where} \qquad \bar{u} = b/a \quad .$$

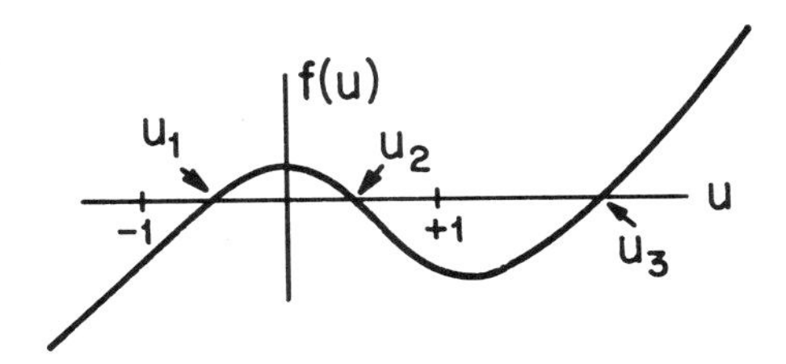

Fig. 7.2. Plot of function f(u) that determines the nutational motion.

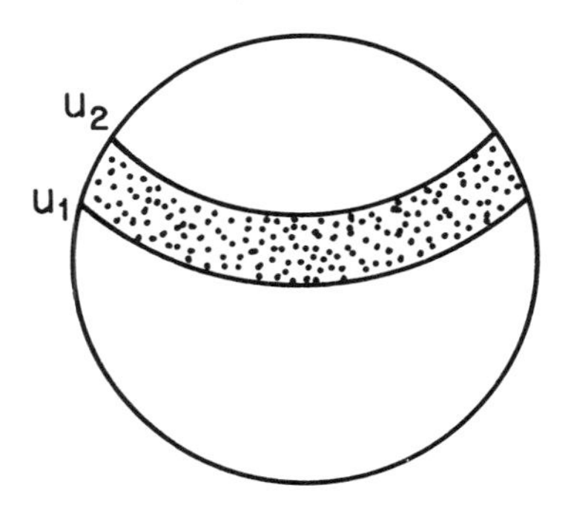

Fig. 7.3. Zone of nutation

There are the following cases (Fig. 7.4):

a) $\bar{u} > u_2 > u_1$ then $\dot{\alpha} > 0$

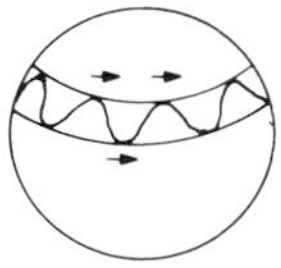

b) $u_2 > \bar{u} > u_1$ then $\dot{\alpha}(2) < 0$
while $\dot{\alpha}(1) > 0$

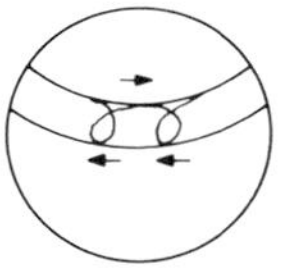

c) $u_2 = \bar{u}$ then $\dot{\alpha}(2) = 0$.

(a)

(b)

(c)

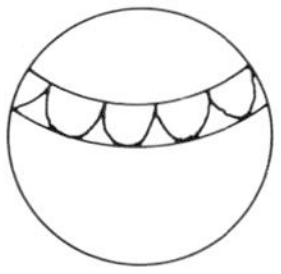

Fig. 7.4. Types of nutational motion.

In case (a), $\dot{\alpha}$ is always of the same sign and therefore the top precesses always in the same direction as it simultaneously nutates. In case (b), the sign of $\dot{\alpha}$ is opposite at u_2 and

u_1 and therefore the precessional motion reverses during the nutation. In case (c), the zero of $\dot{\alpha}$ occurs at u_2. Therefore the precessional motion vanishes, and there is a cusp at u_2. If the spinning top is released from rest, the cusp is required as a boundary condition. The top then begins to nutate by first falling and then rising as shown in the figure.

4.8 THE ROTATION GROUP IN QUANTUM MECHANICS

Let A_μ be a vector operator. Then

$$A'_\mu = \sum R_{\mu\nu} A_\nu \quad . \tag{8.1}$$

In the state space of quantum mechanics A_μ and A'_μ are matrices related by a unitary transformation. Then

$$A'_\mu = U^{-1} A_\mu U \tag{8.2}$$

or

$$U^{-1} A_\mu U = \sum R_{\mu\nu} A_\nu \quad . \tag{8.3}$$

The preceding equation is like (3.17), namely,

$$U^{-1} \sigma_\mu U = \sum R_{\mu\nu} \sigma_\nu$$

where $R^{-1} = R^T$ and σ_μ and U are 2×2 matrices. On the other hand U and A_μ in (8.3) may be of any dimensionality at all.

One may calculate matrix elements as follows from (8.3):

$$\langle m|U^{-1} A_\mu U|n\rangle = \sum R_{\mu\nu} \langle m|A_\nu|n\rangle \quad . \tag{8.3a}$$

Define

$$|n'\rangle = U|n\rangle$$
$$|m'\rangle = U|m\rangle \quad . \tag{8.4}$$

Then

$$\langle m'| = \langle m|U^{-1}$$

and (8.3a) becomes

$$\langle m'|A_\mu|n'\rangle = \sum R_{\mu\nu} \langle m|A_\nu|n\rangle \quad . \tag{8.5}$$

Therefore the matrix elements transform like components of a vector. Equation (8.5) is a Schrödinger type of relation since the operator A_μ is the same for the primed and unprimed frames while the state functions $|n\rangle$ and $|m\rangle$ are transformed. On the other hand, (8.3) is a relation of the Heisenberg kind, since the operator rather than the state function is changed in going from one frame to the other.

The obvious generalization of (8.3) is

$$U^{-1} A_{\mu\nu} \ldots U = \sum R_{\mu\lambda} R_{\nu\rho} \ldots A_{\lambda\rho} \ldots \tag{8.6}$$

where $A_{\mu\nu} \ldots$ is a tensor operator.

4.9 ANGULAR MOMENTUM AND COMMUTATION RULES

The commutation rules of an arbitrary tensor with the angular momentum are determined by its transformation law under the rotation group. These rules are simply calculated by substituting an infinitesimal rotation in (8.6). Let the angular momentum be defined in terms of the generator of an infinitesimal rotation as follows:

$$U = 1 + \frac{1}{i\hbar} \underset{\sim}{\varepsilon} \underset{\sim}{J} \tag{9.1a}$$

where the three dimensional rotation matrix is

$$R_{\mu\nu} = \delta_{\mu\nu} + \varepsilon_{\mu\nu} \quad . \tag{9.1b}$$

Then

$$(1 - \frac{1}{i\hbar} \underset{\sim}{\varepsilon} \underset{\sim}{J})\, A_\mu \,(1 + \frac{1}{i\hbar} \underset{\sim}{\varepsilon} \underset{\sim}{J}) = \sum (\delta_{\mu\nu} + \varepsilon_{\mu\nu})\, A_\nu$$

$$\frac{1}{i\hbar} (A_\mu, \underset{\sim}{J}) \underset{\sim}{\varepsilon} = \sum \varepsilon_{\mu\nu}\, A_\nu \quad .$$

If ε_{12} is an infinitesimal rotation in the 12-plane, then

$$\underset{\sim}{\varepsilon} = (0,0,\varepsilon)$$

and

$$\varepsilon_{12} = -\varepsilon \quad .$$

Therefore

$$(A_1, J_3)\varepsilon = -\varepsilon\ i\hbar\ A_2$$

or

$$(A_1, J_3) = -i\hbar\ A_2 \quad .$$

Therefore, every vector must satisfy the following commutation rules

$$(A_m, J_n) = i\hbar\ \varepsilon_{mnp}\ A_p \quad . \tag{9.2}$$

In particular if $\underset{\sim}{A} = \underset{\sim}{J}$

$$(J_m, J_n) = i\hbar\ \varepsilon_{mnp}\ J_p \tag{9.3}$$

Similarly if A_{mn} is a second rank tensor

$$(A_{mn}, J_p) = i\hbar\ (\varepsilon_{mps} A_{sn} + \varepsilon_{npt} A_{mt}) \quad , \tag{9.4}$$

etc.

We may go back to the classical Poisson brackets by the usual rule $(\ ,\ )/i\hbar \to [\ ,\ ]$.

Finally, we examine the commutation rules of the angular momentum in the body frame. Denote the unit vectors of the body frame by $(\hat{k}_1, \hat{k}_2, \hat{k}_3)$ and the unit vectors of the fixed inertial frame by $(\hat{f}_1, \hat{f}_2, \hat{f}_3)$. Then

$$\sum_s \hat{k}_s K_s = \sum_s \hat{f}_s L_s \quad , \tag{9.5}$$

where K_s and L_s are the components of the same vector operator in the two different frames. Then

$$K_m = \sum (\hat{k}_m \hat{f}_s) L_s = \sum (R^{-1})_{ms} \, L_s \tag{9.6}$$

$$L_m = \sum (\hat{f}_m \hat{k}_s) K_s = \sum R_{ms} K_s \tag{9.7}$$

where

$$R_{ms} = \hat{f}_m \hat{k}_s \tag{9.8}$$

and

$$(R^{-1})_{ms} = (\hat{k}_m \hat{f}_s) \quad .$$

$\underset{\sim}{K}$ and $\underset{\sim}{L}$ by definition generate the same relative rotation of the body frame with respect to the inertial frame. We shall fix the operator properties of $\underset{\sim}{K}$ and $\underset{\sim}{L}$ by assuming that $\underset{\sim}{K}$ rotates $\hat{k}_s$ and leaves $\hat{f}_s$ fixed while $\underset{\sim}{L}$ rotates $\hat{f}_s$ and leaves $\hat{k}_s$ fixed. Therefore, $\underset{\sim}{K}$ changes the components of a fixed vector with respect to the body frame, while $\underset{\sim}{L}$ does the same with respect to the inertial frame. To give the same relative rotation $\underset{\sim}{K}$ and $\underset{\sim}{L}$ must turn their respective frames in opposite directions.

We therefore introduce the following algebra:

$$(\hat{k}_m, \hat{k}_n) = (\hat{f}_m, \hat{f}_n) = (\hat{k}_m, \hat{f}_n) = 0 \tag{9.9a}$$

$$(L_m, \hat{k}_n) = (K_m, \hat{f}_n) = 0 \tag{9.9b}$$

$$(\hat{f}_m, L_n) = i\hbar\, \varepsilon_{mnp} \hat{f}_p \tag{9.9c}$$

$$(\hat{k}_m, K_n) = -i\hbar\, \varepsilon_{mnp} \hat{k}_p \quad . \tag{9.9d}$$

Then

$$(R_{ms}, L_r) = i\hbar\, \varepsilon_{mrt} R_{ts} \tag{9.10}$$

since R_{ms} $(= \hat{f}_m \hat{k}_s)$ is the m-component of $\hat{k}_s$ in the f-system, and

$$(R_{ms}, K_r) = -i\hbar\, \varepsilon_{srt} R_{mt} \tag{9.11}$$

since R_{ms} is also the s-component of $\hat{f}_m$ in the k-system. Then $\underset{\sim}{L}$ works on the first index and $\underset{\sim}{K}$ on the second index of R_{ms}. The sign is opposite in (9.9) and (9.10) since $\underset{\sim}{K}$ and $\underset{\sim}{L}$ must rotate their frames in opposite directions to give the same relative rotation. All other commutation relations are determined by equations (9.7), (9.8), and (9.9), which imply (9.10) and (9.11) and must be regarded as defining $\underset{\sim}{K}$. For by (9.7)

$$(L_m, L_n) = \sum_s (R_{ms}, L_n) K_s + \sum_s R_{ms} (K_s, L_n) \quad .$$

By (9.9) and (9.7)

$$i\hbar \sum_p \varepsilon_{mnp} L_p = i\hbar \sum \varepsilon_{mnp} R_{ps} K_s + \sum R_{ms} (K_s, L_n)$$

$$= i\hbar \sum \varepsilon_{mnp} L_p + \sum R_{ms} (K_s, L_n) \quad .$$

Therefore

$$(K_s, L_n) = 0 \quad . \tag{9.12}$$

Again by (9.7)

$$(L_m, K_n) = \sum (R_{ms}, K_n) K_s + \sum R_{ms} (K_s, K_n) \quad .$$

According to (9.12) the left side vanishes and therefore by (9.11)

$$0 = -i\hbar \sum \varepsilon_{snp} R_{mp} K_s + \sum R_{ms} (K_s, K_n)$$

$$0 = \sum R_{mp} [(K_p, K_n) - i\hbar \, \varepsilon_{snp} K_s]$$

or

$$(K_p, K_n) = -i\hbar \, \varepsilon_{pns} K_s \quad . \tag{9.13}$$

All of the commutation relations just obtained may of course be replaced by Poisson bracket relations in the usual way.

Finally notice that (9.7) takes a hermitian $\underset{\sim}{K}$ into a hermitian $\underset{\sim}{L}$ in view of the commutation relations between R_{mn} and K_n.

4.10 HEISENBERG (EULERIAN) EQUATIONS OF A FREE RIGID BODY

The commutation rules just obtained for the components of the angular momentum with respect to the body and inertial frames are

$$(L_m, L_n) = i\hbar\, \varepsilon_{mnp} L_p \tag{10.1}$$

$$(K_m, K_n) = -i\hbar\, \varepsilon_{mnp} K_p \tag{10.2}$$

$$(L_m, K_n) = 0 \quad . \tag{10.3}$$

Let the Hamiltonian of the rigid body in the body frame be

$$H = \frac{1}{2}\left(\frac{K_1{}^2}{I_1} + \frac{K_2{}^2}{I_2} + \frac{K_3{}^2}{I_3}\right) \quad . \tag{10.4}$$

Then the Heisenberg equations of motion are

$$i\hbar\, \dot{K}_1 = (K_1, H) = \frac{1}{2}\frac{1}{I_2}(K_1, K_2{}^2) + \frac{1}{2}\frac{1}{I_3}(K_1, K_3{}^2) \quad . \tag{10.5}$$

By (10.2)

$$(K_1,K_2^{\ 2}) = K_2(K_1,K_2) + (K_1,K_2)K_2$$

$$= -i\hbar\ (K_2K_3 + K_3K_2)$$

and

$$(K_1,K_3^{\ 2}) = i\hbar\ (K_2K_3 + K_3K_2) \quad .$$

Therefore

$$i\hbar\ \dot{K}_1 = -\frac{i\hbar}{2I_2}(K_3K_2 + K_2K_3) + \frac{i\hbar}{2I_3}(K_2K_3 + K_3K_2)$$

or

$$\dot{K}_1 = \frac{1}{2}\left(-\frac{1}{I_2} + \frac{1}{I_3}\right)(K_2K_3 + K_3K_2)$$

$$\dot{K}_2 = \frac{1}{2}\left(-\frac{1}{I_3} + \frac{1}{I_1}\right)(K_3K_1 + K_1K_3)$$

$$\dot{K}_3 = \frac{1}{2}\left(-\frac{1}{I_1} + \frac{1}{I_2}\right)(K_1K_2 + K_2K_1) \quad . \tag{10.6}$$

These are the Eulerian equations in their quantum form. In the classical limit

$$\langle K_2K_3 + K_3K_2\rangle \to 2\ \langle K_2\rangle\ \langle K_3\rangle = 2\ \langle K_3\rangle\ \langle K_2\rangle \tag{10.7}$$

or

$$<\dot{K}_1> = \left(\frac{1}{I_3} - \frac{1}{I_2}\right) <K_2> <K_3> \tag{10.8}$$

in agreement with (5.29) and the relation

$$<K_s> = I_s \omega_s \quad . \tag{10.9}$$

Similarly one finds by (10.3) and (10.4)

$$\dot{L}_k = 0 \tag{10.10}$$

which expresses the conservation of angular momentum in the inertial frame.

4.11 THE SCHRÖDINGER EQUATION FOR A SYMMETRIC TOP

The time independent equation is

$$H|> = E|> \quad . \tag{11.1}$$

With the Hamiltonian of (10.4) one obtains

$$\frac{1}{2}\left[\frac{K_1{}^2}{I_1} + \frac{K_2{}^2}{I_2} + \frac{K_3{}^2}{I_3}\right] |> = E|> \quad . \tag{11.2}$$

In classical theory the problem of the asymmetric top involves elliptic functions if the three moments are unequal. In the quantum case the result is still more complicated.[13]

Let us consider the case of the symmetric top, namely,

$$I_1 = I_2 = I_\perp$$

$$I_3 \neq I_\perp \quad .$$

Then

$$\frac{1}{2}\left[\frac{K^2}{I_\perp} + K_3{}^2\left(\frac{1}{I_3} - \frac{1}{I_\perp}\right)\right]|\rangle = E|\rangle \quad . \tag{11.3}$$

The following commutation rules are satisfied

$$(L_1,L_2) = i\hbar\, L_3$$

$$(K_1,K_2) = -i\hbar\, K_3$$

$$(L_r,K_s) = 0 \tag{11.4}$$

$$(H,K^2) = (H,K_3) = 0 \quad \text{and} \quad (H,L_s) = 0 \quad . \tag{11.5}$$

The commuting integrals of the motion are then K^2 $(= L^2)$, K_3 and L_3. This is a maximal commuting set. Let the common eigenfunction be $|kmm'\rangle$. Then

$$K^2\,|kmm'\rangle = k(k+1)\,\hbar^2\,|kmm'\rangle \tag{11.6}$$

$$K_3\,|kmm'\rangle = \hbar\, m|kmm'\rangle \tag{11.7}$$

$$L_3\,|kmm'\rangle = \hbar\, m'|kmm'\rangle \quad . \tag{11.8}$$

Let us next introduce a continuous Eulerian basis and let

$\langle\alpha\beta\gamma|k\ell m\rangle = \psi^k_{\ell m}(\alpha\beta\gamma)$. Then K^2, K_3, and L_3 become differential operators and the corresponding differential equations are[14]

$$\left\{-\frac{\partial^2}{\partial\beta^2} - (\cot\beta)\frac{\partial}{\partial\beta} - \frac{1}{\sin^2\beta}\left(\frac{\partial^2}{\partial\alpha^2} + \frac{\partial^2}{\partial\gamma^2} - 2\cos\beta\frac{\partial^2}{\partial\alpha\,\partial\gamma}\right)\right\}$$

$$\times\ \psi^k_{mm'}(\alpha\beta\gamma) = k(k+1)\ \psi^k_{mm'}(\alpha\beta\gamma) \tag{11.9}$$

$$\frac{1}{i}\frac{\partial}{\partial\alpha}\ \psi^k_{mm'}(\alpha\beta\gamma) = m\ \psi^k_{mm'}(\alpha\beta\gamma) \tag{11.10}$$

$$\frac{1}{i}\frac{\partial}{\partial\gamma}\ \psi^k_{mm'}(\alpha\beta\gamma) = m'\ \psi^k_{mm'}(\alpha\beta\gamma) \quad . \tag{11.11}$$

The solution of these equations is known to be[5]

$$\psi^k_{mm'}(\alpha\beta\gamma) = D^k_{mm'}(\alpha\beta\gamma) \tag{11.12}$$

where the $D^k_{mm'}(\alpha\beta\gamma)$ are the matrix elements of the irreducible representations of the rotation group and are known as the Wigner functions. The probability that the top has an orientation $(\alpha\beta\gamma)$ is then

$$\sim |D^k_{mm'}(\alpha\beta\gamma)|^2 \quad .$$

The spectrum is given by (11.3)

$$E(k,m,m') = \frac{1}{2}\left[\frac{k(k+1)\,\hbar^2}{I_\perp{}^2} + m^2\hbar^2\left(\frac{1}{I_3} - \frac{1}{I_\perp}\right)\right] \quad . \tag{11.13}$$

This spectrum has the degeneracy 2k + 1.

The spherical top has the same wave functions and the spectrum is now

$$E(k,m,m') = \frac{k(k+1)\hbar^2}{2I} \quad . \tag{11.14}$$

The spectrum now has the degeneracy $(2k+1)^2$ since E is independent of m.

Equations (11.13) and (11.14) give the rotational spectrum of a molecule.

Correspondence Limit of Symmetric Top:

For the time dependence of $(m|K_\perp|m+1)$ we have

$$(m|K_\perp|m+1) \sim \exp[(i/\hbar)(E_m - E_{m+1})t] \tag{11.15}$$

where

$$E_{m+1} - E_m = \frac{(2m+1)\hbar^2}{2}\left(\frac{1}{I_3} - \frac{1}{I_\perp}\right) \quad . \tag{11.16}$$

In the limit of large m, Eq. (11.16) becomes

$$\left|\frac{E_{m+1} - E_m}{\hbar}\right| \to \left|\frac{1}{I_3} - \frac{1}{I_\perp}\right| \hbar m = \left|\frac{1}{I_3} - \frac{1}{I_\perp}\right| K_3 \quad . \tag{11.17}$$

Equation (11.17) agrees with the classical result (6.5) for

the precessional frequency:

$$\omega_{\text{classical}} = \left| \frac{1}{I_3} - \frac{1}{I_\perp} \right| K_3 \quad . \tag{11.18}$$

Solution of the Time-Dependent Equation:

The general solution may be expressed in terms of

$$\langle\alpha_2\beta_2\gamma_2t_2|\alpha_1\beta_1\gamma_1t_1\rangle = \langle 2|1\rangle \tag{11.19}$$

where $(\alpha_1,\beta_1,\gamma_1)$ describes the orientation of the top at t_1 and $(\alpha_2\beta_2\gamma_2)$ describes its orientation at the later time t_2. Then $|\langle 2|1\rangle|^2$ is the probability of finding the top with the orientation $(\alpha_2\beta_2\gamma_2)$ at time t_2 if it is prepared at the initial time t_1 with the orientation $(\alpha_1\beta_1\gamma_1)$. One may evaluate $\langle 2|1\rangle$ as follows:

$$\langle 2|1\rangle = \sum_{jmm'} \langle 2|jmm'\rangle \langle jmm'|1\rangle \tag{11.20}$$

where we have put in a complete set of intermediate states.

But

$$\langle\alpha_2\beta_2\gamma_2t_2|jmm'\rangle = N_j\, D^j_{mm'}(\alpha_2\beta_2\gamma_2) \exp[-\tfrac{i}{\hbar} E(j,m)t_2] \tag{11.21}$$

where N_j is a normalization factor. Therefore

$$<2|1> = \sum_{jmm'} N_j^{\,2} \left(\exp[-\frac{i}{\hbar} E(j,m)(t_2-t_1)]\right)$$

$$\times D^j_{mm'}(\alpha_2\beta_2\gamma_2)\, D^j_{mm'}(\alpha_1\beta_1\gamma_1)^* \quad . \qquad (11.22)$$

In particular, if the top is spherical

$$<2|1> = \sum_j N_j^{\,2} \exp[-\frac{i}{\hbar} E(j)(t_2-t_1)] \sum_{mm'} D^j_{mm'}(R_2)\, D^j_{mm'}(R_1)^* \qquad (11.23)$$

where R is the rotation corresponding to $(\alpha\beta\gamma)$. Then

$$\sum_{mm'} D^j_{mm'}(R_2)\, D^j_{mm'}(R_1)^* = \sum D^j_{mm'}(R_2)\, D^j_{m'm}(R_1^{\,-1})$$

$$= \sum D^j_{mm}\,(R_2R_1^{\,-1}) \quad . \qquad (11.24)$$

The trace of a representation is called its character, i.e.,

$$\chi^j(R) = \sum_m D^j_{mm}(R) \quad . \qquad (11.25)$$

Therefore one has

$$<2|1> = \sum_j N_j^{\,2} \left(\exp[-\frac{i}{\hbar}\frac{j(j+1)\hbar^2}{2I}(t_2-t_1)]\right) \chi^j(R_2R_1^{\,-1}) \qquad (11.26)$$

for a spherical top. The sum is closely related to a Jacobian theta function.[15]

4.12 ADDITION OF ANGULAR MOMENTA

Consider two tops, A and B, with angular momenta $\underset{\sim}{J}_A$ and $\underset{\sim}{J}_B$. The total angular momentum is $\underset{\sim}{J}_A + \underset{\sim}{J}_B$ according to classical theory. The same statement holds in quantum theory if $\underset{\sim}{J}_A$ and $\underset{\sim}{J}_B$ are interpreted as operators.

The wave equation of a composite system (A + B) is

$$(H_A + H_B)\psi = -\frac{\hbar}{i}\frac{\partial\psi}{\partial t} \tag{12.1}$$

if there is no interaction between the parts A and B, and the wave function is

$$\psi_{A+B} = \psi_A\psi_B \quad . \tag{12.2}$$

In the case of a pair of non-interacting symmetric tops referred to the same center

$$\psi = D^{J_A}_{M_A M_A'}(\alpha_A\beta_A\gamma_A)\, D^{J_B}_{M_B M_B'}(\alpha_B\beta_B\gamma_B) \tag{12.3}$$

and $|\psi(\alpha_A\beta_A\gamma_A\ \alpha_B\beta_B\gamma_B)|^2$ is the probability that the two independent and concentric tops will have the specified orientations for the indicated states of angular momentum.

More generally and concisely the state function of a non-interacting composite system is

$$|A+B\rangle = |A\rangle\ |B\rangle \quad . \tag{12.4}$$

If the states are labeled by angular momentum quantum numbers, then

$$|J_A M_A\ J_B M_B\rangle = |J_A M_A\rangle\ |J_B M_B\rangle \quad . \tag{12.5}$$

If there is interaction between the subsystems, these same functions still provide a complete basis, and the matrix elements of the vector operator $\underset{\sim}{J}_A + \underset{\sim}{J}_B$ between states of the total system are

$$\begin{aligned} &\langle J_A'M_A'\ J_B'M_B'|\underset{\sim}{J}_A+\underset{\sim}{J}_B|J_A''M_A''\ J_B''M_B''\rangle \\ &\quad = \langle J_A'M_A'|J_A|J_A''M_A''\rangle\ \langle J_B'M_B'|1|J_B''M_B''\rangle \\ &\quad\quad + \langle J_A'M_A'|1|J_A''M_A''\rangle\ \langle J_B'M_B'|J_B|J_B''M_B''\rangle \quad . \end{aligned} \tag{12.6}$$

We may abbreviate by writing the total angular momentum vector matrix as

$$\underset{\sim}{J}_A \otimes 1 + 1 \otimes \underset{\sim}{J}_B \tag{12.7}$$

where the product $A \times B$ [16] means

$$(A \otimes B)_{\alpha a;\beta b} = A_{\alpha\beta} B_{ab} \quad . \tag{12.8}$$

The main point now is that the product function $|A\rangle\ |B\rangle$

is not an eigenstate of (angular momentum)^2, even though $|A\rangle$ and $|B\rangle$ are. An equivalent statement is that $\underset{\sim}{J}_A$ and $\underset{\sim}{J}_B$ are irreducible but $J_A \otimes 1 \oplus 1 \otimes J_B$ is reducible.

In fact one may find an S such that

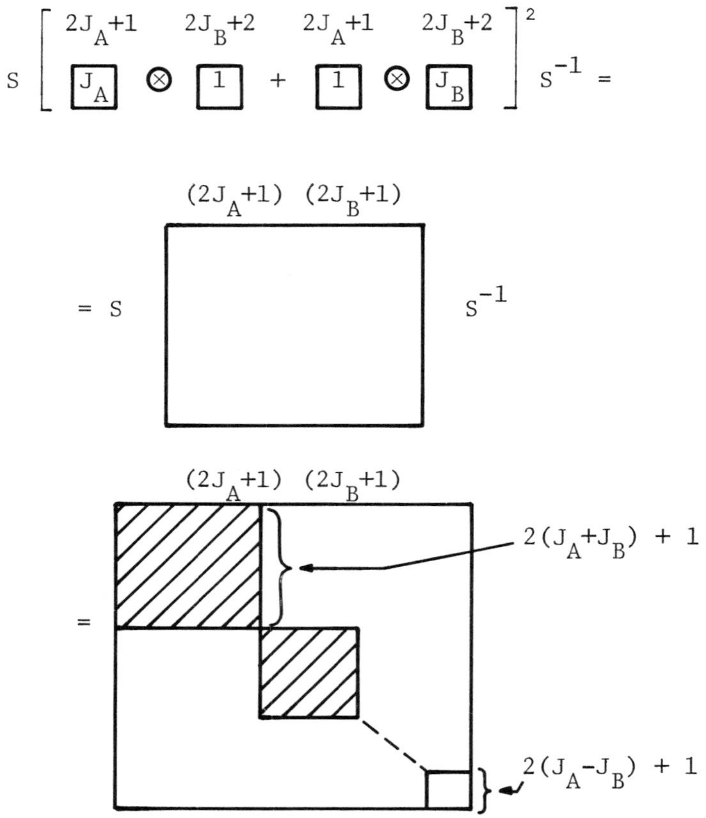

$$S \left[\underset{2J_A+1}{\boxed{J_A}} \otimes \underset{2J_B+2}{\boxed{1}} + \underset{2J_A+1}{\boxed{1}} \otimes \underset{2J_B+2}{\boxed{J_B}} \right]^2 S^{-1} =$$

$$= S \underset{(2J_A+1)\,(2J_B+1)}{\boxed{}} S^{-1}$$

$$= \quad (12.9)$$

where the right side of this equation is the sum of irreducible representations running from $J_A - J_B$ to $J_A + J_B$ in steps of $\hbar$.

The dimensions of the various matrices are indicated in this equation. Frequently one abbreviates by writing

$$D^{J_A} \otimes D^{J_B} = D^{J_A - J_B} \oplus \cdots \oplus D^{J_A + J_B} \qquad . \qquad (12.10)$$

On the left one has an outer product and on the right one has an outer sum in the sense displayed by equation (12.9). For example, one has for two electrons, each of spin $\frac{1}{2}\hbar$,

$$S \left[\overset{2}{\boxed{\vec{S}}} \otimes \overset{2}{\boxed{1}} + \overset{2}{\boxed{1}} \otimes \overset{2}{\boxed{\vec{S}}} \right]^2 S^{-1}$$

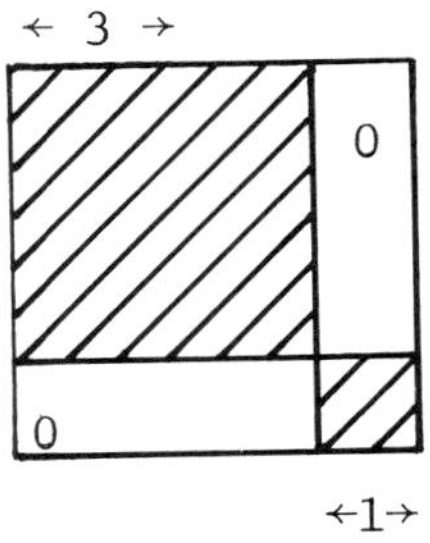

$$= \begin{pmatrix} \underbrace{\blacksquare}_{3\times 3} & 0 \\ 0 & \underbrace{\blacksquare}_{1\times 1} \end{pmatrix} \qquad (12.11)$$

or

$$D^{1/2} \otimes D^{1/2} = D^1 \oplus D^0 \qquad .$$

In general one may check the dimensions of the two sides as follows:

$$(2J_A+1)\ (2J_B+1) = \sum_{|J_A-J_B|}^{J_A+J_B} (2J+1) \quad . \tag{12.12}$$

Here the right side is the total dimensionality of the unreduced matrix and J increases by steps of one between (J_A-J_B) and J_A+J_B.

The preceding result is the basis for the quantum mechanical rule for the addition of angular momenta: the magnitude of the quantum resultant of $\underset{\sim}{J}_A$ and $\underset{\sim}{J}_B$ runs from (J_A-J_B) to J_A+J_B and therefore has the same limits as the classical sum. However, the quantum resultant progresses in steps of $\hbar$ whereas the corresponding classical sum runs continuously. Clearly if the angular momentum is large compared to $\hbar$, then the discrete quantum sequence will simulate the continuous classical distribution.

The preceding rule for the addition of angular momenta is of great practical utility in many applications of quantum mechanics, such as the vector model of the atom, for example. In this particular model it is recognized that the ith electron has two kinds of angular momentum, namely orbital $(\underset{\sim}{\ell}_i)$ and spin $(\underset{\sim}{s}_i)$. The total orbital and the total spin angular momenta are

$$\underset{\sim}{L} = \sum \underset{\sim}{\ell}_i$$

$$\underset{\sim}{S} = \sum \underset{\sim}{s}_i$$

respectively, and the sum of these is

$$\underset{\sim}{J} = \underset{\sim}{L} + \underset{\sim}{S} \quad .$$

For light atoms it may be shown to a first approximation that $\underset{\sim}{L}$ and $\underset{\sim}{S}$ are approximately conserved and, of course, $\underset{\sim}{J}$, the total angular momentum of the atom, is exactly conserved if the atom is free. By compounding $\underset{\sim}{L}$ and $\underset{\sim}{S}$ according to the quantum law of vector additions, one obtains the LS-classification of atomic states ${}^{2S+1}L_J$, where S and L are approximate and where J is an exact quantum number.

We have described the rotation of a planet and a molecule and indicated how these descriptions are bridged by the correspondence principle. The procedure for combining angular momenta in an atom has just been described and one can see how this same recipe leads to the corresponding classical law for the addition of angular momenta. In the next chapter planetary and atomic motions will be further investigated and one will be able to see how the correspondence principle works out in more detail.

Although the underlying theory therefore permits one to describe these extremely different physical systems in the framework of a single formalism, the intuitive pictures and the observational procedures that are important for the solar system on the one hand and for the molecule or the atom on the other, are quite different. In the case of the solar system we follow the motion in space and time without sacrificing complementary information about momentum and energy. On the other hand, for the molecule or atom we observe energy levels (spectra) and transition amplitudes (line intensities) and do not follow the motion in space and time. The existence of macroscopic quantum systems like superfluids permits one to obtain both of these complementary kinds of observational information about truly coherent states, subject to the limitations of the uncertainty principle; but unfortunately not enough is presently known about these systems.

NOTES ON CHAPTER 4

1. If group multiplication is commutative, then the group is called abelian; otherwise non-abelian. The three-dimensional rotation group is non-abelian.
2. This two dimensional representation, U(R), of the rotation group is necessarily irreducible, for if it were reducible, then every matrix of the reduced representation would be of the form $\begin{pmatrix} a & 0 \\ 0 & b \end{pmatrix}$. These matrices would all mutually commute and therefore so would the U(R). It is

easy to check, however, that the U(R) do not commute.

3. Notice that

$$e^{UAU^{-1}} = 1 + UAU^{-1} + \frac{1}{2}(UAU^{-1})(UAU^{-1}) + \cdots$$

$$= U(1 + A + \frac{A^2}{2} + \cdots)U^{-1} = U\, e^A\, U^{-1} \quad .$$

4. Equation (3.32) may be proved as follows. Since the z and the initial z' axes are the same, we have first

$$R_{z'}(\alpha) = R_z(\alpha) \tag{N4.1}$$

Next one tilts by the angle β about the new y'-axis. This rotation is $R_{y'}(\beta)$ and may be obtained by first rotating by the angle β about the old (fixed) y-axis and then rotating the old y-axis into the position of the new y-axis. Therefore

$$R_{y'}(\beta) = R_z(\alpha)\, R_y(\beta)\, R_z(\alpha)^{-1} \quad . \tag{N4.2}$$

The right hand side of this equation may be interpreted as the result of moving the axis (y) of $R_y(\beta)$ into the new axis (y') by making a rotating α about z. [Compare with the interpretation of (3.28) as a rotated rotation.]

Finally the rotation of γ about the new z-axis is

$$R_{z'}(\gamma) = R_{y'}(\beta)[R_z(\alpha)\, R_z(\gamma)\, R_z(\alpha)^{-1}]\, R_{y'}(\beta)^{-1} \ . \tag{N4.3}$$

The right hand side now describes a rotation of γ about z, then a rotation of the z-axis first of α about z and then of β about y'.

Since rotations about the same axis commute, (N4.3) becoems simply

$$R_{z'}(\gamma) = R_{y'}(\beta)\, R_z(\gamma)\, R_{y'}(\beta)^{-1} \tag{N4.4}$$

and therefore by (N4.1)

$$\begin{aligned} R_{z'}(\gamma)\, R_{y'}(\beta)\, R_{z'}(\alpha) &= [R_{y'}(\beta)\, R_z(\gamma)\, R_{y'}(\beta)^{-1}][R_{y'}(\beta)][R_z(\alpha)] \\ &= R_{y'}(\beta)\, R_z(\alpha)\, R_z(\gamma) \end{aligned} \tag{N4.5}$$

and finally by (N4.2)

$$R_{z'}(\gamma)\, R_{y'}(\beta)\, R_{z'}(\alpha) = R_z(\alpha)\, R_y(\beta)\, R_z(\gamma) \quad .$$

This last relation is (3.32) and asserts the equivalence of the sequence $\alpha\beta\gamma$ with respect to the fixed axes with the reverse sequence with respect to the moving axes.

Eulerian angles are frequently defined with respect to the rotating frame and labeled (ϕ,θ,ψ). For other differences in the definitions used by different authors, see Goldstein, op. cit. in Chapter 1, page 108.

We have here adopted the same conventions for $D(\alpha,\beta,\gamma)$ as Edmonds, reference 1 of this chapter, page 55. The formulas for angular velocity in Edmonds, page 66, agree with (4.19) of this chapter.

5. See, e.g., references 1 and 2 of this chapter. There is one and only one irreducible representation for every dimensionality. Therefore $J = 0, 1/2, 1, \cdots$.
6. See Appendix C for additional discussion of connections between vector analysis and spin algebra; also problems 1 and 2.
7. See, for example, Goldstein, op. cit.
8. Equations (4.19) agree with Edmonds, reference 1, p. 66.
9. See Goldstein, page 163, op. cit.
10. For brief discussion see Landau and Lifshitz, op. cit., p. 116. For detailed mathematics, see Whittaker, op. cit., p. 144. For additional discussion of elliptic functions, see reference 5 of this chapter.
11. See, e.g., Whittaker, Landau, and Goldstein, op. cit.
12. Our discussion of the top will be very abbreviated. For more details see Goldstein and other references given there, in particular reference 3 of this chapter.
13. C. von Winter, Physica 20, 274 (1954).
14. See Appendix D and exercise 16 for two different derivations of this equation.

15. See exercises 18 and 19.

16. $A \otimes B$ is the outer product.

BIBLIOGRAPHY FOR CHAPTER 4

To the earlier references we add:

1. A. R. Edmonds, Angular Momentum in Quantum Mechanics, Princeton Univ. Press, 1957, page 55.

2. E. P. Wigner, Group Theory and Its Application to the Quantum Mechanics of Atomic Spectra, Academic Press, 1959.

3. F. Klein and A. Sommerfeld, Theorie des Kreisels.

4. H. B. G. Casimir, Rotation of a Rigid Body in Quantum Mechanics, Thesis, Leiden, 1931.

For additional discussion of elliptic functions see

5. E. T. Whittaker and G. N. Watson, A Course of Modern Analysis, Cambridge, 1940.

PROBLEMS ON CHAPTER 4

1. Let $A = \underset{\sim}{a}\underset{\sim}{\sigma}$, $B = \underset{\sim}{b}\underset{\sim}{\sigma}$, $C = \underset{\sim}{c}\underset{\sim}{\sigma}$. Prove

$$\left((A,B),C\right) = -4\ (\underset{\sim}{a} \times \underset{\sim}{b}) \times \underset{\sim}{c} \cdot \underset{\sim}{\sigma}$$

$$\mathrm{Tr}\ ABC = i(\underset{\sim}{a} \times \underset{\sim}{b}) \cdot \underset{\sim}{c} \qquad .$$

2. Prove the matrix relation

$$\big((A,B),C\big) = A\{B,C\} - \{A,C\}B - B\{A,C\} + \{B,C\}A$$

and derive from it the corresponding vector relation

$$(\underset{\sim}{a} \times \underset{\sim}{b}) \times \underset{\sim}{c} = \underset{\sim}{b}(\underset{\sim}{a}\underset{\sim}{c}) - (\underset{\sim}{b}\underset{\sim}{c})\underset{\sim}{a} \quad .$$

Here { } means anticommutator.

3. Show that

$$(G_k, G_\ell) = i\,\varepsilon_{k\ell m}\, G_m$$

has a 3 × 3 solution:

$$(G_k)_{mn} = i\,\varepsilon_{mkn} \quad .$$

4. Check Table 3.1.

5. From (4.13) prove

$$\overset{\circ}{\omega}_k = \dot{\alpha}\delta_{k3} + \dot{\beta}\, R_{k2}(-\alpha,0,0) + \dot{\gamma}\, R_{k3}(-\alpha,-\beta,0)$$

and verify relations (4.15).

6. Show that the angle of rotation w is related to the Eulerian angles by the formula

$$\cos\frac{w}{2} = \cos\frac{\beta}{2}\cos\frac{\alpha+\gamma}{2} \quad .$$

[Compare (3.31) with (3.27) and take the trace.]

7. By applying (4.11) to the radius vector $\underset{\sim}{r}$, find the transformation law for the velocity of a particle when one goes from an inertial to a rotating frame. By applying (4.11) a second time, obtain the corresponding transformation law for the acceleration. Identify the centrifugal and the Coriolis terms.

8. Show that

$$\dot{R}_{km} R^{-1}_{ms} = - \varepsilon_{ksm}\omega_m \quad .$$

9. Prove

$$\frac{dB_k}{dt} + \sum \varepsilon_{k\ell m}\bar{\omega}_\ell B_m = \sum R^{-1}_{km}\dot{A}_m$$

where B_R are the components in the body frame, and
A_R are the components in the inertial frame,
by using result of preceding problem.

10. Show that matrices with the property $R^{-1} = R^T$ (T = transposed) form a group under multiplication.

11. Show that the torque is independent of the position of the origin, if the total force vanishes.

12. Start from

$$L_m = \sum I_{mn}\omega_n \quad , \qquad L_s = \sum R_{st}K_t \quad .$$

Show that

$$K_s = \sum \bar{I}_{sm} \bar{\omega}_m$$

where $\bar{\omega}_m$ and $\bar{I}_{mn}$ are found in the body frame according to the following expressions:

$$\omega_m = \sum R_{mn} \bar{\omega}_n \quad , \qquad I_{st} = \sum R_{sm} \bar{I}_{mn} R^{-1}_{nt} \quad .$$

By calculating dK_s/dt, show that

$$\sum \bar{I}_{sm} \frac{d\bar{\omega}_m}{dt} + \varepsilon_{smn} \bar{\omega}_m \bar{I}_{np} \bar{\omega}_p = 0$$

with the aid of problem 8.

13. Write the Poisson brackets of the angular momentum with the rotation matrix that connects the body and inertial frames.

14. Find commutator (A_{mn}, L_p) when A_{mn} is second rank tensor.

15. Show how Lagrange's equations written for a free rigid body reduce to Euler's equations.

16. Calculate for a symmetric top:

$$p_\alpha = \frac{\partial L}{\partial \dot{\alpha}} , \qquad p_\beta = \frac{\partial L}{\partial \dot{\beta}} , \qquad p_\gamma = \frac{\partial L}{\partial \dot{\gamma}} \quad .$$

Express $\underset{\sim}{K}$ in terms of Eulerian angles and conjugate momenta by inverting preceding equations for $\dot{\alpha}$, $\dot{\beta}$, and $\dot{\gamma}$

and substituting in

$$K_1 = I\,\omega_1 = I\,(-\dot{\alpha}\,\sin\beta\,\cos\gamma + \dot{\beta}\,\sin\gamma)$$

....

Show that

$$p_\alpha = \frac{\hbar}{i}\frac{\partial}{\partial\alpha}\,, \qquad p_\gamma = \frac{\hbar}{i}\frac{\partial}{\partial\gamma}\,, \qquad \text{and} \qquad p_\beta = \frac{\hbar}{i}\left(\frac{\partial}{\partial\beta} + \frac{1}{2}\cot\beta\right)$$

and finally that

$$K^2 = L^2 = -\hbar^2\left\{\frac{\partial^2}{\partial\beta^2} + \cot\beta\,\frac{\partial}{\partial\beta} + \frac{1}{\sin^2\beta}\left(\frac{\partial^2}{\partial\alpha^2} + \frac{\partial}{\partial\gamma^2} - 2\cos\beta\,\frac{\partial^2}{\partial\alpha\partial\gamma}\right)\right\}\,.$$

Write the Hamilton-Jacobi and Schrödinger equation of the symmetric top.

Notice that the volume element is $\sin\beta\, d\alpha\, d\beta\, d\gamma$ and therefore p_β will not be hermitian without the additional term.

17. A solid sphere is made into a fixed-point top by supporting it at one point on its surface. The top is set spinning vertically ($\beta = 0$) with angular velocity ω.

  a) Find the minimum ω such that the top is stable in a vertical position.

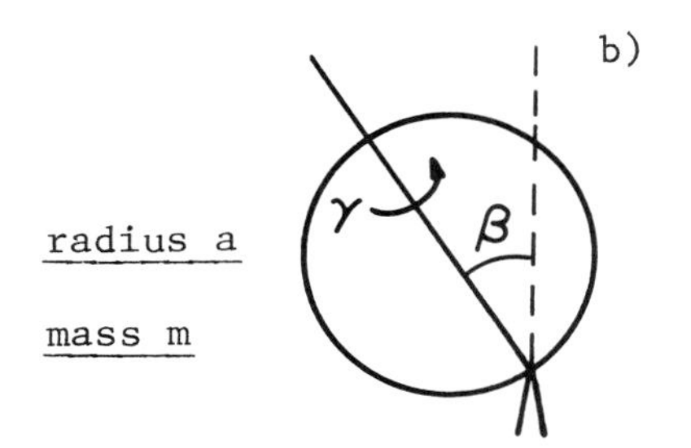

b) For ω less than this value, find the maximum β to which the top will nutate if released from a vertical position.

18. The character $\chi^j(R)$ defined in (11.15) depends only on the magnitude of the rotation. Prove

$$\chi^j(R) = \frac{\sin\left(j + \frac{1}{2}\right)\omega}{\sin\frac{1}{2}\omega}$$

by using (3.40), where ω is the angle of the rotation.

19. By using the result of the previous exercise, investigate the relation between (11.16) and the Jacobian theta function. (See Whittaker and Watson, op. cit.)

20. Calculate

$$\frac{\hbar}{2}\left(\underset{\sim}{\sigma}_A \otimes 1 + 1 \otimes \underset{\sim}{\sigma}_B\right) = \underset{\sim}{S} \qquad .$$

By performing similarity transformation on $\underset{\sim}{S}^2$, show how triplet and singlet representations arise.

21. Use the solution of problem 3 to calculate $D^1(\alpha,\beta,\gamma)$. A particle of spin $\hbar$ is prepared in a state with the z-component of its spin equal to zero. What is the probability of finding the z-component of its spin to be $\hbar$ in

the $\hat{n}$ direction?

22. The Brownian motion of the orientation of a rigid body has been calculated in the isotropic case with the result

$$F(\phi)\,d\phi = \frac{2}{\pi}\sin\frac{\phi}{2}\sum_{n=0}^{\infty}(2n+1)\sin\left(n+\frac{1}{2}\right)\phi\; e^{-n(n+1)Dt}$$

where $F(\phi)\,d\phi$ is the probability that the final angle of orientation is in the interval $(\phi, \phi+d\phi)$. W. H. Furry, Phys. Rev. 107, 7 (1957).

Discuss relation of this result to problem 19.

23. Specify the orientation of a rigid body by the generalized coordinates q^{α}, $\alpha = 1,2,3$. Show that the Lagrangian may be written $L = \frac{1}{2} g_{\alpha\beta}\dot{q}^{\alpha}\dot{q}^{\beta}$ where $g_{\alpha\beta} = \lambda_{\alpha}{}^{k}(-)\, I_{k\ell}\, \lambda_{\beta}{}^{\ell}(-)$. Here $I_{k\ell}$ is the inertia tensor in the body frame and $\lambda_{\alpha}{}^{k}(-)$ is defined in Appendix (E.8). Show that the Hamiltonian and angular momentum in the body frame can be written $H = \frac{1}{2} g^{\alpha\beta} p_{\alpha}p_{\beta}$ and $K_{\gamma} = -I_{rs}\,\lambda^{s\alpha}(-)\, p_{\alpha}$ where $\lambda_{k}{}^{\alpha}(-) = g^{\alpha\beta}\lambda_{k\beta}(-)$ and $p_{\alpha} = \partial L/\partial\dot{q}_{\alpha}$. Verify (10.1), (10.2), and (10.3) by using the canonical commutation rules of the p's and q's.

CHAPTER 5

PLANETARY AND ATOMIC SYSTEMS

5.1 INTRODUCTION

A solar system is similar to an atom in the sense that it is organized by the attraction of a very massive sun or nucleus. Then to a first approximation every planet or electron moves in the field of the central body.[1] This first approximation may next be corrected by determining the perturbation of these orbits by either the Newtonian attraction between the planets or the Coulomb repulsion of the electrons.

Here we are not concerned with detailed refinements of the perturbation method but only with the first approximation which is determined essentially by the Kepler problem. This simple picture of non-interacting electrons in a central Coulomb field leads to shell structure and the magic numbers ($2n^2$) of the periodic table, if one takes into account the

Pauli exclusion principle and the doubling of states caused by the spin.

The relation of the rotation group to the dynamics of a rigid body has been discussed in the previous chapter. The relation of this group to the classification of atomic states (by the angular momentum quantum number) is also clear in principle and is the basis for atomic spectroscopy. The relation of the rotation group to the radial Kepler problem is, however, not so generally appreciated and will be discussed in the present chapter. From this viewpoint the structure of the periodic system is also determined by the rotation group.

5.2 SEPARATION OF CENTER OF MASS MOTION

An isolated system is characterized by the following constants of the motion: $\underset{\sim}{P}$, $\underset{\sim}{L}$, and $\underset{\sim}{G}$. From the constancy of $\underset{\sim}{P}$ and $\underset{\sim}{G}$ one infers uniform motion of the center of mass:

$$\underset{\sim}{X} = (\underset{\sim}{P}/M)t + \underset{\sim}{G}/M \quad . \tag{2.1}$$

Let the radius vector of the ith particle be

$$\underset{\sim}{X} + \underset{\sim}{x}_i \quad . \tag{2.2}$$

Then the kinetic energy is

$$T = \sum \frac{1}{2} m_i (\dot{\underset{\sim}{X}} + \dot{\underset{\sim}{x}}_i)^2$$

$$= \sum \frac{1}{2} m_i (\dot{\underset{\sim}{X}}^2 + 2\dot{\underset{\sim}{X}}\ \dot{\underset{\sim}{x}}_i + \dot{\underset{\sim}{x}}_i{}^2) \quad . \tag{2.3}$$

But

$$\sum m_i (\underset{\sim}{X} + \underset{\sim}{x}_i) = M\underset{\sim}{X}$$

by definition of $\underset{\sim}{X}$. Then

$$\sum m_i \underset{\sim}{x}_i = 0 \tag{2.4}$$

and therefore

$$T = \frac{1}{2} M\dot{\underset{\sim}{X}}^2 + \sum \frac{1}{2} m_i \dot{\underset{\sim}{x}}_i{}^2 \quad . \tag{2.5}$$

Likewise

$$\underset{\sim}{L} = \sum (\underset{\sim}{X} + \underset{\sim}{x}_i) \times \underset{\sim}{p}_i$$

$$= \sum \underset{\sim}{x}_i \times \underset{\sim}{p}_i + \underset{\sim}{X} \times \underset{\sim}{P} \quad . \tag{2.6}$$

Let us next specialize these formulas to the two body problem. Then the Lagrangian is

$$L = \frac{1}{2} M\dot{\underset{\sim}{X}}^2 + \sum \frac{1}{2} m_i \dot{\underset{\sim}{x}}_i{}^2 - V(\underset{\sim}{x}_1 - \underset{\sim}{x}_2) \tag{2.7}$$

subject to

$$m_1 \underset{\sim}{x}_1 + m_2 \underset{\sim}{x}_2 = 0 \quad . \tag{2.8}$$

Introduce the relative coordinates:

$$\underset{\sim}{x} = \underset{\sim}{x}_1 - \underset{\sim}{x}_2 \quad . \tag{2.9}$$

Then

$$\dot{\underset{\sim}{x}} = \dot{\underset{\sim}{x}}_1 - \dot{\underset{\sim}{x}}_2 \tag{2.10}$$

$$= \dot{\underset{\sim}{x}}_1 \left(1 + (m_1/m_2)\right) \tag{2.11a}$$

or

$$= - \dot{\underset{\sim}{x}}_2 \left(1 + (m_2/m_1)\right) \tag{2.11b}$$

by (2.8). Therefore

$$\frac{1}{2} m_1 \dot{\underset{\sim}{x}}_1{}^2 + \frac{1}{2} m_2 \dot{\underset{\sim}{x}}_2{}^2 = \frac{1}{2} \left[\left(\frac{m_2}{m_1+m_2}\right)^2 m_1 + \left(\frac{m_1}{m_1+m_2}\right)^2 m_2 \right] \dot{\underset{\sim}{x}}^2$$

$$= \frac{1}{2} \frac{m_1 m_2}{m_1+m_2} \dot{\underset{\sim}{x}}^2 \quad . \tag{2.12}$$

Define the reduced mass:

$$\mu = \frac{m_1 m_2}{m_1+m_2} \quad . \tag{2.13}$$

Then

$$\frac{1}{2} m_1 \dot{\underset{\sim}{x}}_1{}^2 + \frac{1}{2} m_2 \dot{\underset{\sim}{x}}_2{}^2 = \frac{1}{2} \mu \dot{\underset{\sim}{x}}^2 \tag{2.14}$$

and the Lagrangian becomes

$$L = \frac{1}{2} M\dot{\underset{\sim}{X}}^2 + \frac{1}{2} \mu\dot{\underset{\sim}{x}}^2 - V(\underset{\sim}{x}) \quad . \tag{2.15}$$

The canonical momenta are now the total momentum, which is conjugate to the center of mass coordinates

$$P_\alpha = \frac{\partial L}{\partial \dot{X}_\alpha} = M\dot{X}_\alpha \tag{2.16a}$$

and the internal momentum which is conjugate to the relative coordinates:

$$p_\alpha = \frac{\partial L}{\partial \dot{x}_\alpha} = \mu\dot{x}_\alpha \tag{2.16b}$$

and therefore the Hamiltonian is

$$H = \frac{\underset{\sim}{P}^2}{2M} + \frac{\underset{\sim}{p}^2}{2\mu} + V(\underset{\sim}{x}) \tag{2.17}$$

or

$$H = H_{ext} + H_{int} \tag{2.18}$$

where

$$H_{ext} = \underset{\sim}{P}^2/2M \tag{2.18a}$$

and

$$H_{int} = \underset{\sim}{p}^2/2\mu + V(x) \quad . \tag{2.18b}$$

The total angular momentum is by (2.6)

$$\underset{\sim}{L} = \underset{\sim}{X} \times \underset{\sim}{P} + \underset{\sim}{x}_1 \times \underset{\sim}{p}_1 + \underset{\sim}{x}_2 \times \underset{\sim}{p}_2 \quad . \tag{2.19}$$

One may choose a coordinate system (center of mass) for which the external momentum vanishes:

$$\underset{\sim}{P} = \underset{\sim}{p}_1 + \underset{\sim}{p}_2 = 0 \quad . \tag{2.20}$$

Then

$$H = \frac{\underset{\sim}{p}^2}{2\mu} + V(x) \tag{2.21}$$

$$\underset{\sim}{L} = \underset{\sim}{x} \times \underset{\sim}{p}_1 \quad . \tag{2.22}$$

But

$$\mu^{-1}\, \underset{\sim}{p} = m_1^{-1}\, \underset{\sim}{p}_1 - m_2^{-1}\, \underset{\sim}{p}_2 \tag{2.23}$$

by (2.16b) and (2.10). Therefore $\underset{\sim}{p} = \underset{\sim}{p}_1$ if (2.20) holds, and consequently

$$\underset{\sim}{L} = \underset{\sim}{x} \times \underset{\sim}{p} \tag{2.24}$$

where $\underset{\sim}{p}$ is the internal momentum that also determines the internal kinetic energy in (2.21).

For the discussion of the internal problem we shall need only equations (2.21) and (2.24) for H and $\underset{\sim}{L}$ respectively.

5.3 CENTRAL FORCE PROBLEM (CLASSICAL)

Let us examine the tensors which may be formed from $\underset{\sim}{L}$, $\underset{\sim}{p}$, and $\underset{\sim}{r}$ (which we shall now use instead of $\underset{\sim}{x}$). The scalar products vanish

$$\underset{\sim}{L} \cdot \underset{\sim}{r} = \underset{\sim}{L} \cdot \underset{\sim}{p} = 0 \quad . \tag{3.1}$$

The vector products are

$$\underset{\sim}{M} = \underset{\sim}{L} \times \underset{\sim}{r} \tag{3.2}$$

$$\underset{\sim}{\mathcal{L}} = \underset{\sim}{L} \times \underset{\sim}{p} \tag{3.3}$$

and are simply related:

$$\mu \frac{d\underset{\sim}{M}}{dt} = \underset{\sim}{\mathcal{L}} \quad . \tag{3.4}$$

Next consider

$$\frac{d\underset{\sim}{\mathcal{L}}}{dt} = \underset{\sim}{L} \times \frac{d\underset{\sim}{p}}{dt} = \underset{\sim}{L} \times \underset{\sim}{F} \quad . \tag{3.5}$$

If F is associated with a central potential, then

$$\underset{\sim}{F} = - \left(\frac{dV}{dr}\right) (\underset{\sim}{r}/r) \tag{3.6}$$

and

$$\frac{d\underset{\sim}{L}}{dt} = \left(-\frac{1}{r}\frac{dV}{dr}\right)(\underset{\sim}{L} \times \underset{\sim}{r}) \tag{3.7a}$$

$$= \left(-\frac{1}{r}\frac{dV}{dr}\right)\underset{\sim}{M} \quad . \tag{3.7b}$$

$\underset{\sim}{M}$ in turn may be expressed in terms of $d\hat{\underset{\sim}{r}}/dt$ where

$$\hat{\underset{\sim}{r}} = \underset{\sim}{r}/r \tag{3.8}$$

for

$$\underset{\sim}{M} = (\underset{\sim}{r} \times \underset{\sim}{p}) \times \underset{\sim}{r} = r^2\underset{\sim}{p} - \underset{\sim}{r}(\underset{\sim}{r}\underset{\sim}{p}) \quad . \tag{3.9}$$

On the other hand

$$\frac{d\hat{\underset{\sim}{r}}}{dt} = \frac{d}{dt}(\underset{\sim}{r}/r) = \frac{1}{r}\dot{\underset{\sim}{r}} + \underset{\sim}{r}\left(-\frac{1}{r^2}\frac{dr}{dt}\right) \quad . \tag{3.10}$$

By (3.9) and (3.10)

$$\underset{\sim}{M} = \mu r^3 \frac{d\hat{\underset{\sim}{r}}}{dt} \quad . \tag{3.11}$$

Finally by (3.7b)

$$\frac{d\underset{\sim}{L}}{dt} = -\mu r^2 \frac{dV}{dr}\frac{d\hat{\underset{\sim}{r}}}{dt} \quad . \tag{3.12}$$

There are two particularly simple cases:

$$\text{a)} \qquad V = -\frac{k}{r} \qquad \text{Kepler} \tag{3.13}$$

$$\text{b)} \qquad V = kr^2 \qquad \text{Harmonic Oscillator} \quad . \tag{3.14}$$

In case a)

$$\frac{d}{dt}\underset{\sim}{L} = -\mu k \frac{d}{dt}\underset{\sim}{\hat{r}}$$

or

$$\frac{d}{dt}(\underset{\sim}{L} + \mu k\,\underset{\sim}{\hat{r}}) = 0 \quad . \tag{3.15}$$

In case b) by (3.7b)

$$\frac{d\underset{\sim}{L}}{dt} = -2k\,\underset{\sim}{M}$$

or

$$\frac{d^2\underset{\sim}{L}}{dt^2} = -\frac{2k}{\mu}\underset{\sim}{L} \quad . \tag{3.16}$$

We shall discuss just the Kepler case. Then define the Runge-Lenz vector:

$$\underset{\sim}{A} = \underset{\sim}{L} + k\mu\,\underset{\sim}{\hat{r}} = (\underset{\sim}{r} \times \underset{\sim}{p}) \times \underset{\sim}{p} + k\mu\,\underset{\sim}{\hat{r}} \quad . \tag{3.17}$$

This is now an integral of the motion by (3.15).

Since $\underset{\sim}{L}$ is perpendicular to the plane of the orbit, it is clear that $\underset{\sim}{A}$ lies in this plane

$$\underset{\sim}{A}\,\underset{\sim}{L} = 0 \qquad . \tag{3.18}$$

The constant vector $\underset{\sim}{A}$ provides a second vector integral of the motion and yields the equation of the orbit as follows:

$$\underset{\sim}{A}\,\underset{\sim}{r} = \underset{\sim}{\mathcal{L}}\,\underset{\sim}{r} + k\mu\, r \qquad . \tag{3.19}$$

But

$$\underset{\sim}{\mathcal{L}}\,\underset{\sim}{r} = (\underset{\sim}{L} \times \underset{\sim}{p})\underset{\sim}{r} = \underset{\sim}{L}(\underset{\sim}{p} \times \underset{\sim}{r}) = -\,\underset{\sim}{L}^2 = -\,L^2 \qquad . \tag{3.20}$$

Therefore

$$A\ r\ \cos\theta = -\ L^2 + k\mu\ r$$

or

$$\frac{1}{r} = \frac{k\mu}{L^2}\,(1 - \varepsilon\cos\theta) \tag{3.21}$$

where θ is the polar angle between $\underset{\sim}{r}$ and $\underset{\sim}{A}$, and where

$$\varepsilon = \frac{A}{\mu k} \qquad . \tag{3.22}$$

(3.21) is the equation of the orbit and describes a conic with eccentricity ε.

The magnitude of $\underset{\sim}{A}$ is therefore related to the eccentricity of the conic by (3.22). The direction of $\underset{\sim}{A}$ also determines the orientation of the conic as follows. Choose the z-axis along the direction of the angular momentum. Then

$$A_z = 0 \qquad (3.23a)$$

$$A_x = -\, Lp_y + k\mu \frac{x}{r} \qquad (3.23b)$$

$$A_y = Lp_x + k\mu \frac{y}{r} \quad . \qquad (3.23c)$$

Choose the x-axis along the major axis of the conic. Notice that when the orbit crosses the x-axis

$$p_x = 0 \ , \qquad y = 0 \quad .$$

Then

$$A_y = 0 \qquad (3.24)$$

and therefore $\underset{\sim}{A}$ is directed along the major axis at this point in the orbit. However $\underset{\sim}{A}$ is constant and it is therefore always directed along the major axis.

In summary, $\underset{\sim}{L}$ is perpendicular to the plane of the orbit, while $\underset{\sim}{A}$ gives the orientation of the conic in this plane and its magnitude determines the eccentricity of this conic.

Finally the eccentricity may be determined in terms of the energy and angular momentum as follows:

$$A^2 = L^2 \underset{\sim}{p}^2 + \frac{2k\mu L}{r} (-xp_y + yp_x) + \mu^2 k^2$$

$$= L^2\underset{\sim}{p}^2 + \frac{2k\mu L}{r}(-L) + \mu^2k^2$$

$$= 2\mu L^2\left(\frac{\underset{\sim}{p}^2}{2\mu} - \frac{k}{r}\right) + \mu^2k^2$$

$$A^2 = 2\mu L^2E + \mu^2k^2 \tag{3.25}$$

and

$$\varepsilon = \left(1 + \frac{2L^2E}{\mu k^2}\right)^{1/2} \quad . \tag{3.26}$$

One has the following classification

$\varepsilon > 1$	$E > 0$	hyperbola
$\varepsilon = 1$	$E = 0$	parabola
$\varepsilon < 1$	$E < 0$	ellipse
$\varepsilon = 0$	$E = -\dfrac{\mu k^2}{2L^2}$	circle .

The existence of the integral of the motion, $\underset{\sim}{A}$, is evidently the essence of the Kepler problem and any other way of discussing the problem must represent some transcription of this fact. Nevertheless, it is interesting to study this motion, since it is so fundamental, from other points of view and to compare it with motion in other kinds of central potential.

5.4 MOTION IN A GENERAL POTENTIAL

In order to gain a more general point of view, consider motion in a potential which is static but not necessarily spherically symmetric. We should also like to see how the simplicity of the Kepler problem appears in this wider context.

We may introduce polar coordinates. Then the kinetic energy is

$$T = \frac{\mu}{2} (\dot{r}^2 + r^2 \sin^2\theta \, \dot{\phi}^2 + r^2\dot{\theta}^2) \tag{4.1}$$

and

$$p_r = \mu\dot{r} \tag{4.2}$$

$$p_\theta = \mu r^2\dot{\theta} \tag{4.3}$$

$$p_\phi = \mu r^2 \sin^2\theta \, \dot{\phi} \quad . \tag{4.4}$$

Then the Hamiltonian is

$$H = \frac{1}{2\mu} \left[p_r{}^2 + \frac{1}{r^2} p_\theta{}^2 + \frac{1}{r^2 \sin^2\theta} p_\phi{}^2 \right] + V(r,\theta,\phi) \quad . \tag{4.5}$$

The Hamilton-Jacobi equation is now

$$\frac{1}{2\mu}\left[\left(\frac{\partial W}{\partial r}\right)^2 + \frac{1}{r^2}\left(\frac{\partial W}{\partial \theta}\right)^2 + \frac{1}{r^2 \sin^2\theta}\left(\frac{\partial W}{\partial \phi}\right)^2\right] + V(r,\theta,\phi) = \alpha_1 \tag{4.6}$$

since the potential is static.[2]

Let

$$W(r,\theta,\phi) = W_r(r) + W_\theta(\theta) + W_\phi(\phi) \quad . \tag{4.7}$$

Then

$$\frac{1}{2\mu}\left[(W'_r)^2 + \frac{1}{r^2}(W'_\theta)^2 + \frac{1}{r^2 \sin^2\theta}(W'_\phi)^2\right] + V(r,\theta,\phi) = \alpha_1 \quad . \tag{4.8}$$

In general the Hamilton-Jacobi equation is not separated by the ansatz (4.7). However, if V does not depend on ϕ, then the only function in the above equation which does depend on ϕ is $(W'_\phi)^2$. But since W'_ϕ is also equal, by the same equation, to a function of r and θ only, W'_ϕ must be independent of ϕ, as well as of r and θ. Therefore

$$W'_\phi = \alpha_\phi \tag{4.9}$$

where α_ϕ is a constant, i.e., independent of r, θ, and ϕ.

One may next put (4.9) back into (4.8) to obtain

$$\frac{1}{2\mu}\left[(W'_r)^2 + \frac{1}{r^2}(W'_\theta)^2 + \frac{1}{r^2\sin^2\theta}\alpha_\phi^2\right] + V(r,\theta) = \alpha_1 \quad . \tag{4.10}$$

Again if $V(r,\theta)$ depends on θ, the separation cannot be continued in general. On the other hand, if V is spherically symmetric, one may now make the same argument about the function

$$(W'_\theta)^2 + \frac{1}{\sin^2\theta}\alpha_\phi^2$$

that was just made about W'_ϕ to obtain (4.9). Therefore one now has

$$(W'_\theta)^2 + \frac{1}{\sin^2\theta}\alpha_\phi^2 = \alpha_\theta^2 \tag{4.11}$$

or

$$\frac{dW}{d\theta} = \left[\alpha_\theta^2 - \frac{1}{\sin^2\theta}\alpha_\phi^2\right]^{1/2} \tag{4.11a}$$

where α_θ is a second constant of integration.

Finally, putting (4.11) back into (4.10), one gets

$$(W'_r)^2 + \frac{\alpha_\theta^2}{r^2} = 2\mu[\alpha_1 - V(r)]$$

or

$$\frac{dW_r}{dr} = \left[2\mu \, [\alpha_1 - V(r)] - \frac{\alpha_\theta^{\,2}}{r^2} \right]^{1/2} \quad . \tag{4.12}$$

Therefore if the potential is central, the Hamilton-Jacobi equation is completely separable in spherical coordinates.

The solution of the dynamical problem is then reduced to certain quadratures.

Choose α_ϕ (the z-component of the angular momentum) so that

$$\alpha_\phi = 0 \quad . \tag{4.13}$$

Then

$$W(\theta) = \alpha_\theta \theta \tag{4.14}$$

$$W(r) = \int \left[2\mu \, [\alpha_1 - V(r)] - \frac{\alpha_\theta^{\,2}}{r^2} \right]^{1/2} dr \tag{4.15}$$

and finally

$$W(r,\theta) = \alpha_\theta \theta + \int \left[2\mu \, [\alpha_1 - V(r)] - \frac{\alpha_\theta^{\,2}}{r^2} \right]^{1/2} dr \quad . \tag{4.16}$$

Since the motion takes place in a plane, Hamilton's characteristic function depends on two variables and two constants of integration:

$$W(r,\theta,\alpha_1,\alpha_\theta) \quad . \tag{4.16a}$$

The Hamilton-Jacobi method now permits one to perform a contact transformation from (q,p) to the initial values of these variables, which are of course constants. One may regard the constants of integration α_1 and α_θ as initial momenta. The corresponding constant coordinates (β_1 and β_θ) are then obtained from the generating function (4.16a) by the formulas (6.8) of Chapter 3, namely,

$$\frac{\partial W}{\partial \alpha_1} = t + \beta_1 \tag{4.17}$$

$$\frac{\partial W}{\partial \alpha_\theta} = \beta_\theta \tag{4.18}$$

or by (4.16)

$$\mu \int \frac{dr}{[2\mu(\alpha_1 - V) - (\alpha_\theta{}^2/r^2)]^{1/2}} = t + \beta_1 \tag{4.19}$$

and

$$\theta - \alpha_\theta \int \frac{dr}{[2\mu(\alpha_1 - V) - (\alpha_\theta{}^2/r^2)]^{1/2}} \frac{1}{r^2} = \beta_\theta \quad . \tag{4.20}$$

Equation (4.20) gives the equation of the orbit while (4.19) connects position in orbit with time. The two constants of integration $(\alpha_1, \alpha_\theta)$ are the energy and angular momentum.

In this way the problem of motion in a general central potential is solved. The solution of course includes the Kepler case but does not reveal its privileged position. To go further, we now turn to the action and angle variables; and in order to describe this method, we shall make a digression from the subject of orbits in the next section.

5.5 ACTION AND ANGLE VARIABLES

The simplest motion which goes beyond the rigid body approximation to a many particle system is obtained by letting the individual particles undergo small oscillations about their equilibrium positions. As long as the displacements from equilibrium are small, the potential energy may be approximated by a quadratic form in these displacements or in other generalized coordinates. Those coordinates which actually reduce the potential and the kinetic energy simultaneously to a sum of squares are known as the normal coordinates and in terms of these the Hamiltonian is

$$H = \sum H_i = \sum \frac{1}{2m_i} (P_i^2 + m_i^2 \omega_i^2 Q_i^2) \tag{5.1}$$

where m_i is an effective mass and the sum is over all degrees of freedom. Each degree of freedom is then represented by a harmonic oscillator of frequency ω_i. The use of normal co-ordinates ensures that these oscillators are uncoupled. One obtains in this way the classical theory of small vibrations or the quantum theory of vibrational spectra.

The classical description of such a system may be visualized in phase space. Then the representative point moves in such a way that its projection on the P_iQ_i-plane traces out an ellipse according to (5.1); for each H_i is constant, since the separate oscillations are uncoupled. In figure (5.1) each ellipse corresponds to a different energy (E_i). If the displacements from equilibrium become larger, then the motion becomes anharmonic and the projections are no longer ellipses.

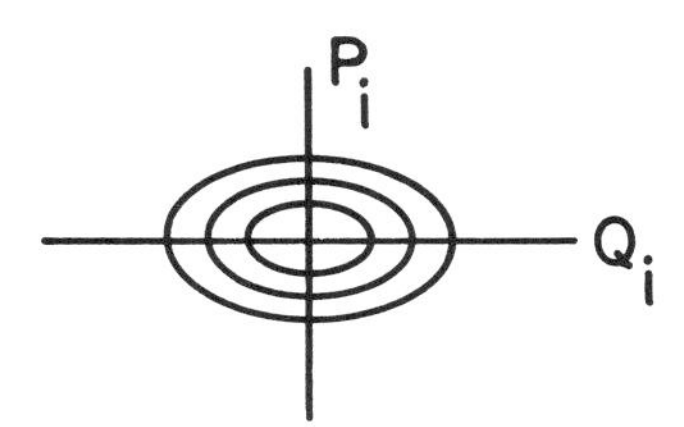

Fig. 5.1. Contours of constant energy for the harmonic oscillator.

If the dynamical system is such that there is more than one equilibrium point, then the phase diagram may resemble figure (5.2). There are then periodic motions (contour C_A) about the equilibrium A and also periodic motions (contour C_B) about the equilibrium B. If the energy is sufficiently large, however, then the motion will be represented by the large contour C which encloses both equilibrium points A and B. Motions of type C are separated from motions like C_A and C_B by the figure-eight curve passing through the origin and known as the separatrix. If the energy is represented as the height above the P_iQ_i-plane, then A and B represent the bottoms of depressions while S is a saddle point and represents an unstable equilibrium.

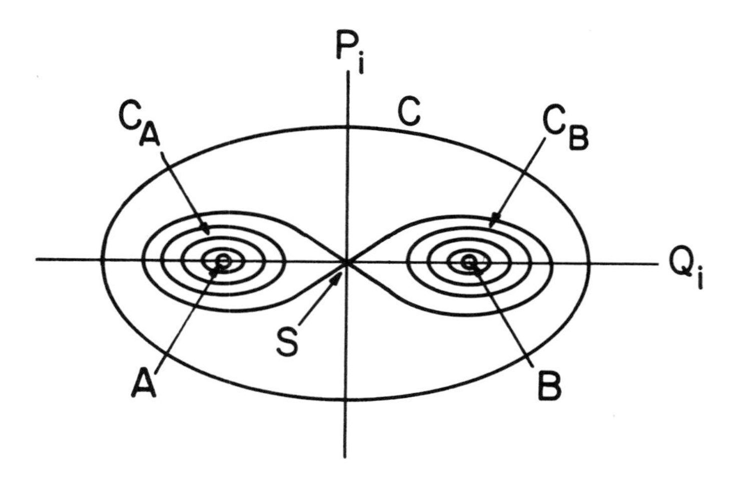

Fig. 5.2. Contours of constant energy when there are two equilibrium positions.

We shall now discuss a general motion that is periodic in each P_iQ_i-plane separately. Such a motion is called multiply or conditionally periodic.

We shall limit our discussion to cases in which one may write

$$W = \sum_i W_i(q_i\ \alpha_1 \cdots \alpha_f) \tag{5.2}$$

where the $\alpha_1 \cdots \alpha_f$ are the constants of integration. These functions do not contain the time. Then

$$p_s = \frac{\partial W}{\partial q_s} = \sum_i \frac{\partial W_i(q_i,\alpha)}{\partial q_s} = \frac{dW_s(q_s,\alpha)}{dq_s} \quad . \tag{5.3}$$

Therefore p_s is entirely determined by q_s unless dW_s/dq_s is not single valued.

Now consider the motion of the system in phase space and its projection on one p_sq_s-plane. Suppose that the motion is limited in configuration space so that

$$q_s^{min} < q_s < q_s^{max} \quad . \tag{5.4}$$

Then the variation of p_s is also restricted by (5.3) and (5.4).

If W_s is of the form $\sqrt{R_s}$, then p_s is a two valued function of q_s. Then the motion after projection onto the

$q_s p_s$-plane is shown in the figure (5.3). The motion in each coordinate plane will be periodic, but the total motion will not be periodic unless the periods are all commensurable.

The points q_{min} and q_{max} are called turning points. One may now define

$$J_s = \oint p_s dq_s \tag{5.5}$$

where the path of integration in the complex q-plane is shown in the figure (5.4). One should integrate from q_{min} to q_{max}

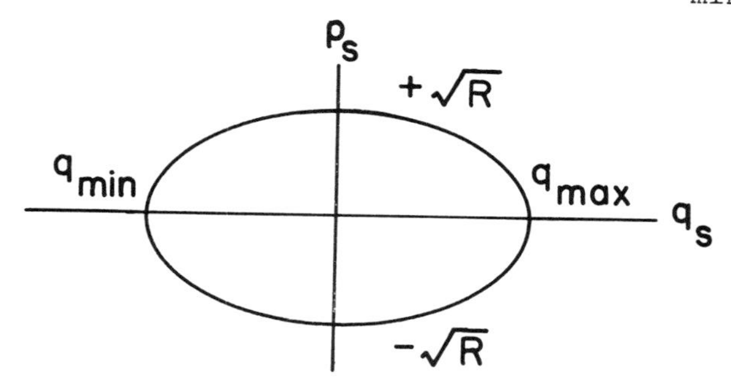

Fig. 5.3. Plot of equation (5.3).

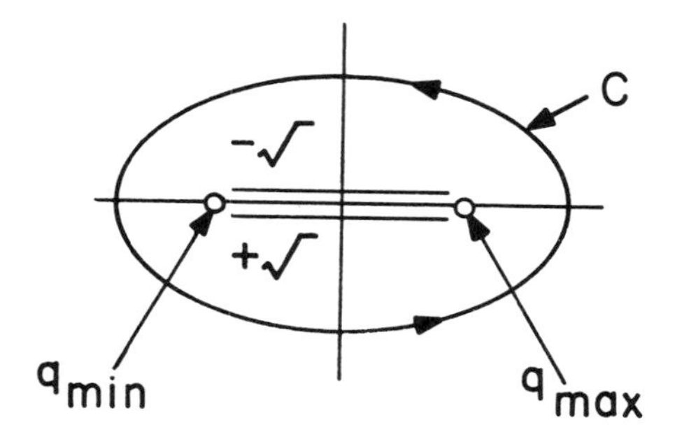

Fig. 5.4. Contour for equation (5.5) in complex q-plane.

with $+\sqrt{R}$ and back with $-\sqrt{R}$, or integrate around the contour (C) shown in figure (5.4). Then by (5.5) and (5.3)

$$J_s = J_s(\alpha_1 \cdots \alpha_f) \quad . \tag{5.6}$$

One may now invert these equations to obtain

$$\alpha_s = \alpha_s(J_1 \cdots J_f) \tag{5.7}$$

and in particular

$$H = \alpha_1 = \alpha_1(J_1 \cdots J_f) \quad . \tag{5.8}$$

The constants of integration (α_s) were first introduced as generalized momenta. The J_s are also momenta and are called action variables.

One may now apply the form of Hamilton-Jacobi theory described in (6.9) - (6.10) of Chapter 3. The coordinates conjugate to the action variables J_k are the so-called angle variables ω_k. Then (6.9b) of that chapter becomes

$$\omega_k = \frac{\partial W}{\partial J_k} \tag{5.9}$$

where Q_k and P_k of (6.9b) become ω_k and J_k, respectively.

By Hamilton's equations one now obtains

$$\frac{d\omega_k}{dt} = \frac{\partial H}{\partial J_k}(J_1 \cdots J_f) = \text{constant} = \nu_k \text{ , say.} \tag{5.10}$$

Then

$$\omega_k = \nu_k t + \gamma_k \quad . \tag{5.11}$$

The ν_k are the frequencies of the projected motions, as one may see in the following way. Let q_m run through a complete cycle while all the other coordinates are fixed. Then

$$\Delta_m \omega_n = \oint \frac{\partial \omega_n}{\partial q_m} dq_m = \oint \frac{\partial}{\partial q_m}\left(\frac{\partial W}{\partial J_n}\right) dq_m = \frac{\partial}{\partial J_n} \oint \frac{\partial W}{\partial q_m} dq_m$$

$$= \frac{\partial}{\partial J_n} \oint p_m dq_m = \frac{\partial}{\partial J_n} J_m = \delta_{nm} \quad . \tag{5.12}$$

Therefore

$$\Delta_n \omega_n = 1 \tag{5.13}$$

and by (5.11)

$$\nu_n \Delta t_n = 1 \tag{5.14}$$

where Δt_n is the time required to complete a cycle in the $q_n p_n$-plane. Therefore ν_n is the corresponding frequency. It follows that the frequencies of a general motion may be found without completely solving the equations of motion if one is able to make use of the formula (5.10):

$$\nu_n = \frac{\partial H}{\partial J_n} \quad . \tag{5.15}$$

Let us finally express the q_i as functions of the ω_i. Then by (5.13) the q_i are periodic functions of the angle variables with period one, and therefore the q_i can be expressed as a multiple Fourier series with the arguments $2\pi\ (n_1\omega_1 + n_2\omega_2 + \cdots)$ where the $(n_1 n_2\ \cdots)$ are arbitrary integers. Then by (5.11) one sees that the entire motion may be represented as a superposition of simple harmonic motions with fundamental frequencies ν_i.

Before concluding this section we should like to mention the fact that the action and angle variables are very closely related to the emission and absorption operators that are so important in many body theory and in field theory.[3]

In the present context, action and angle variables have been of great importance in celestial mechanics and in the old quantum theory.[4] We shall now continue our discussion to show how they are used in the description of orbits.

5.6 MOTION IN A CENTRAL POTENTIAL DESCRIBED BY ACTION AND ANGLE VARIABLES

If there are three degrees of freedom, one expects the following:

$$E = E(J_1, J_2, J_3) \quad . \tag{6.1}$$

However, when the motion takes place in a central potential, it turns out that

$$E = E(J_r, J_\theta + J_\phi) \tag{6.2}$$

where, according to (5.5) and section 4 of this chapter,

$$J_\phi = \oint p_\phi d\phi = \oint \frac{\partial W}{\partial \phi} d\phi = 2\pi\alpha_\phi \tag{6.3}$$

$$J_\theta = \oint p_\theta d\theta = \oint \frac{\partial W}{\partial \theta} d\theta = \oint \left[\alpha_\theta^2 - \frac{\alpha_\theta^2}{\sin^2\theta} \right]^{1/2} d\theta$$

$$= 2\pi(\alpha_\theta - \alpha_\phi) \tag{6.4}$$

$$J_r = \oint p_r dr = \oint \frac{\partial W}{\partial r} dr = \oint \left[2\mu\,\alpha_1 - 2\mu\,V + \frac{\alpha_\theta^2}{r^2} \right]^{1/2} dr \ . \tag{6.5}$$

The assertion (6.2) follows from the equation:

$$J_r = \oint \left[2\mu\, E - 2\mu\, V(r) - \frac{(J_\theta + J_\phi)^2}{4\pi^2 r^2} \right]^{1/2} dr \qquad (6.6)$$

obtained by combining equations (6.3) - (6.5). When V(r) is the Kepler potential, however, the above integration may be carried out with the result

$$E = - \frac{2\pi^2 \mu k^2}{(J_r + J_\theta + J_\phi)^2} \qquad (6.7)$$

$$= E(J_r + J_\theta + J_\phi) \quad . \qquad (6.8)$$

It follows from (5.15) and (6.2) that in the general central potential

$$\nu_\phi = \nu_\theta \neq \nu_r \quad . \qquad (6.9)$$

In the Kepler case, however, (6.8) holds and

$$\nu_\phi = \nu_\theta = \nu_r \quad . \qquad (6.10)$$

That is, motion in the Kepler, central, and noncentral potentials are distinguished by one, two, and three frequencies respectively. The second frequency in the central potential corresponds to a rotation of the orbit in its plane, and the third frequency in the noncentral case corresponds to the rotation of the plane of the orbit.

5.7 DEGENERACY

The motion is said to be m-fold degenerate if there are m-relations of the following type

$$\sum_{s=1}^{f} j_{sk}\nu_s = 0 \qquad k = 1\cdots m \tag{7.1}$$

where the j_{sk} are integers. When there are f-1 such relations, all periods are commensurable, the motion is periodic and the orbit is closed. The motion is then said to be completely degenerate.

In the central force problem there is partial degeneracy

$$\nu_1 = \nu_2 \tag{7.2}$$

but in the Kepler problem there is complete degeneracy

$$\nu_1 = \nu_2 = \nu_3 \quad . \tag{7.3}$$

When the degeneracy is less than complete, it is still useful to transform to the f-m independent frequencies. Therefore define the generating function

$$F = \sum_{s=1}^{f} \sum_{k=1}^{m} j_{sk}\omega_s J_k' + \sum_{k=m+1}^{f} \omega_k J_k' \quad . \tag{7.4}$$

This function describes a contact transformation from the variables (ω, J) to the new variables (ω', J') as follows:

$$\omega_k' = \frac{\partial F}{\partial J_k'} , \qquad J_k = \frac{\partial F}{\partial \omega_k}$$

so that

$$\omega_k' = \sum_{s=1}^{f} j_{sk}\omega_s \qquad k = 1\cdots m \qquad (7.5a)$$

$$\omega_k' = \omega_k \qquad k = m+1\cdots f \qquad (7.5b)$$

and the new action variables are given implicitly by the equations

$$J_s = \sum_{k=1}^{m} J_k' \, j_{sk} + \sum_{k=m+1}^{f} J_k' \, \delta_{ks} \quad . \qquad (7.6)$$

The new frequencies are

$$\nu_k' = \dot{\omega}_k' = \sum_{s=1}^{f} j_{sk}\nu_s = 0 \qquad k = 1\cdots m \qquad (7.7a)$$

$$\nu_k' = \nu_k \qquad k = m+1\cdots n \quad . \qquad (7.7b)$$

Therefore m of the new frequencies vanish and the remaining are independent.

Conditional Periodicity

Even when there is no degeneracy at all, the ratio of the different frequencies may be approximated by the ratio of sufficiently large numbers. For example, one may write

$$N_1\tau_1 \cong N_2\tau_2 \tag{7.8}$$

where τ_1 and τ_2 are two arbitrary periods and N_1 and N_2 are sufficiently large. Then in the very long time $N_1\tau_1$, the motion is periodic to the accuracy of (7.8). In this approximate sense all multiply periodic systems return arbitrarily close to an initial state.

Degeneracy in Quantum Theory

In quantum theory, degeneracy means that there are several states belonging to a given energy level, or that there are several assignments of quantum numbers which correspond to the same energy. On the other hand, if the system is classically degenerate, that is, if the frequencies ν_1 and ν_2 for example are commensurable, then the energy depends on $M_1J_1 + M_2J_2$ where M_1 and M_2 are integers. But according to the Bohr-Sommerfeld quantization condition the possible values of J_s are

$$J_s = N_s h \quad .$$

Therefore, in this example, classical degeneracy implies that the energy depends on $(M_1N_1 + M_2N_2)$ h, and that all choices of N_1 and N_2 leading to the same value of $(M_1N_1 + M_2N_2)$ h correspond to the same energy. Therefore, classical degeneracy implies quantum degeneracy, as the correspondence principle requires.

Separability of Hamilton-Jacobi Equation

It has been shown that the Kepler and oscillator potentials are the only examples for which all motions are simply periodic (completely degenerate). These potentials also have the property that the Hamilton-Jacobi equation is separable in a continuous family of prolate spheroidal coordinate systems with arbitrary orientation of the unique axis and arbitrary interfocal distance, namely,

$$x = c \sinh \xi \sin\eta \cos\phi$$

$$y = c \sinh \xi \sin\eta \sin\phi$$

$$z = c \cosh \xi \cos\eta \quad .$$

This coordinate system reduces in the limits $c = 0$ and $c = \infty$ to spherical and parabolic coordinates. Therefore, the Hamilton-Jacobi equation is separable not only in the two distinct coordinate systems, spherical and parabolic, but also in a continuous family which interpolates between them. It has been shown that the completely degenerate character of the Kepler and oscillator motions may be inferred from the existence of this continuous family of separable coordinates.[5]

5.8 THE HEISENBERG DESCRIPTION OF THE KEPLER PROBLEM AND THE MOTION OF A TOP

The general central field problem is characterized by an angular momentum, $\underset{\sim}{L}$, which is perpendicular to the plane of the orbit. The Kepler problem is additionally distinguished by a second vector integral $\underset{\sim}{A}$:

$$\underset{\sim}{A} = \underset{\sim}{L} \times \underset{\sim}{p} + k\mu\,\hat{\underset{\sim}{r}} \tag{8.1}$$

which determines the direction of the major axis and the eccentricity of the conic.

If one writes for the general central potential

$$V(r) = -\frac{k}{r} + \phi(r) \tag{8.2}$$

then it follows from (3.11) and (3.12) of this chapter that

$$\frac{d}{dt}\underset{\sim}{A} = \left(-\frac{1}{r}\frac{d\phi}{dr}\right)(\underset{\sim}{L} \times \underset{\sim}{r}) \quad . \tag{8.3}$$

Therefore, $\underset{\sim}{A}$ precesses in general according to the preceding equation. Consequently, the general case of motion in a central potential may be pictured in terms of a precessing conic, that also has a changing eccentricity.

It follows that even in the bound case the orbit is not closed except in the Kepler case.

If one studies the same problem with action and angle variables, one finds that motion in a central field is in general characterized by two periods which are not commensurable while Keplerian motion is distinguished by a single period. This statement is equivalent to the remark that in general $\underset{\sim}{A}$ precesses so that the orbit is closed only in the Kepler case.

To examine the group theoretical basis of this situation let us first write down the algebra of the Poisson brackets. We shall see that the algebra of the Kepler problem is then the same as the algebra of a rigid body.

$$[H, L_s] = 0 \tag{8.4}$$

$$[H, A_s] = 0 \tag{8.5}$$

$$[L_s, L_t] = \varepsilon_{stm} L_m \tag{8.6}$$

$$[L_s, A_t] = \varepsilon_{stm} A_m \tag{8.7}$$

$$[A_s, A_t] = -2\mu H \varepsilon_{stm} L_m \quad . \tag{8.8}$$

Equation (8.7) follows from the vector character of $\underset{\sim}{A}$. The bracket in (8.8) may be calculated in a straightforward way:

$$[A_s, A_t] = \sum_m \left[\frac{\partial A_s}{\partial x_m} \frac{\partial A_t}{\partial p_m} - \frac{\partial A_t}{\partial x_m} \frac{\partial A_s}{\partial p_m} \right] \tag{8.9}$$

where

$$A_s = (x^m p_m) p_s - \underset{\sim}{p}^2 x_s + \mu k \, (x_s/r) \quad . \tag{8.10}$$

Next define

$$M_s = (-\, 2\mu \, H)^{-1/2} \, A_s \quad . \tag{8.11}$$

Then

$$[L_s, L_t] = \varepsilon_{stm} L_m \tag{8.12}$$

$$[L_s, M_t] = \varepsilon_{stm} M_m \tag{8.13}$$

$$[M_s, M_t] = \varepsilon_{stm} L_m \quad . \tag{8.14}$$

Equations (8.12) - (8.14) define the classical algebra. One may now go over to the quantum equations in the usual way by replacing the Poisson brackets by commutators.

$$(L_s, L_t) = i\hbar \, \varepsilon_{stm} L_m \tag{8.15}$$

$$(L_s, M_t) = i\hbar \, \varepsilon_{stm} M_m \tag{8.16}$$

$$(M_s, M_t) = i\hbar \, \varepsilon_{stm} L_m \quad . \tag{8.17}$$

Note that the following classical relation is still valid:

$$\underset{\sim}{L}\,\underset{\sim}{A} = \underset{\sim}{L}\,\underset{\sim}{M} = 0 \tag{8.18}$$

but the quantum forms of $\underset{\sim}{A}$ and $\underset{\sim}{A}^2$ are slightly different from the classical expressions since the quantum operators must be hermitian. That is

$$A_s = \mathcal{L}_s + k\mu\, \hat{r}_s \tag{8.19}$$

as before, but

$$\mathcal{L}_s = \frac{1}{2}\, \varepsilon_{smn}(L_m p_n + p_n L_m) \tag{8.20}$$

so that

$$\mathcal{L}_s^+ = \mathcal{L}_s \quad . \tag{8.21}$$

It then follows that

$$\underset{\sim}{A}^2 = (\mu k)^2 + (2\mu\, H)(\underset{\sim}{L}^2 + \hbar^2) \tag{8.22}$$

instead of (3.25). Then by (8.11)

$$\underset{\sim}{M}^2 = \frac{(\mu k)^2}{(-\,2\mu\, H)} - \underset{\sim}{L}^2 - \hbar^2 \quad .$$

If this relation is solved for the Hamiltonian, one has

$$H = -\,\frac{1}{2}\ \frac{\mu k^2}{\underset{\sim}{L}^2 + \underset{\sim}{M}^2 + \hbar^2} \quad . \tag{8.23}$$

To relate these equations to the algebra of a rigid body,

introduce

$$\underset{\sim}{J} = \frac{1}{2} (\underset{\sim}{M} + \underset{\sim}{L}) \tag{8.24}$$

$$\underset{\sim}{K} = \frac{1}{2} (\underset{\sim}{M} - \underset{\sim}{L}) \quad . \tag{8.25}$$

Then

$$\begin{aligned}(J_1, J_2) &= \frac{1}{4} [(M_1,M_2) + (L_1,L_2) + (M_1,L_2) + (L_1,M_2)] \\ &= i\hbar \frac{1}{2} (L_3 + M_3) \\ &= i\hbar J_3 \quad .\end{aligned}$$

Similarly

$$\begin{aligned}(K_1, K_2) &= \frac{1}{4} [(M_1,M_2) + (L_1,L_2) - (M_1,L_2) - (L_1,M_2)] \\ &= i\hbar \frac{1}{2} (L_3 - M_3) \\ &= - i\hbar K_3 \quad .\end{aligned}$$

Finally,

$$\begin{aligned}(K_1, J_2) &= \frac{1}{4} [(M_1,M_2) + (M_1,L_2) - (L_1,M_2) - (L_1,L_2)] \\ &= i\hbar \frac{1}{2} [0 M_3 + 0 L_3] \\ &= 0\end{aligned}$$

or

$$(J_m, J_n) = i\hbar\, \varepsilon_{mnp} J_p \tag{8.26}$$

$$(K_m, K_n) = -\, i\hbar\, \varepsilon_{mnp} K_p \tag{8.27}$$

$$(J_m, K_n) = 0 \quad . \tag{8.28}$$

These are just the equations of a rigid body where J_s and K_s are the components of the angular momentum referred to inertial and body frames, respectively. However, one still has the constraint (8.18) or

$$(\underset{\sim}{J} + \underset{\sim}{K})(\underset{\sim}{J} - \underset{\sim}{K}) = 0 \quad .$$

In view of the commutation rules, this constraint then becomes

$$\underset{\sim}{J}^2 = \underset{\sim}{K}^2 \quad . \tag{8.29}$$

Therefore the algebra (8.26) - (8.28) is subject to this additional restriction for the physical application.

The Hamiltonian is given in terms of $\underset{\sim}{J}$ and $\underset{\sim}{K}$ by (8.23)

$$H = -\frac{1}{2}\ \frac{\mu k^2}{2(\underset{\sim}{J}^2 + \underset{\sim}{K}^2) + \hbar^2}$$

since

$$2(\underset{\sim}{J}^2 + \underset{\sim}{K}^2) = \underset{\sim}{L}^2 + \underset{\sim}{M}^2$$

by (8.24) and (8.25) Again by (8.29)

$$H = -\frac{1}{2} \frac{\mu k^2}{4\underset{\sim}{J}^2 + \hbar^2} \quad . \tag{8.30}$$

But the possible eigenvalues of $\underset{\sim}{J}^2$ are

$$J^2 = j(j + 1) \hbar^2 \tag{8.31}$$

and therefore

$$H = -\frac{1}{2} \frac{\mu k^2}{[4j(j+1) + 1] \hbar^2} \quad .$$

But

$$4j(j+1) + 1 = (2j + 1)^2 \quad .$$

Let

$$N = 2j + 1 \quad .$$

Then

$$H = -\frac{1}{2} \frac{\mu k^2}{\hbar^2} \frac{1}{N^2} \quad . \tag{8.32}$$

This is the Balmer formula since the possible values of N are 1,2,3,$\cdots$ if $j = 0, 1/2, \cdots$.

Except for the connection between the energy and the vectors $\underset{\sim}{J}$ and $\underset{\sim}{K}$, the algebra of this problem is exactly the same as one would write for a rigid body, namely, (8.26) -

(8.28) including the constraint (8.29). Therefore, the complete set of commuting observables may also be chosen to be the same, namely, $J^2 (= K^2)$, J_z, K_z and therefore the eigenfunctions are also formally the same:

$$\underset{\sim}{J}^2 \, D^j_{mm'}(\alpha\beta\gamma) = j(j+1)\hbar^2 \, D^j_{mm'}(\alpha\beta\gamma) \tag{8.33a}$$

$$J_z \, D^j_{mm'}(\alpha\beta\gamma) = m\hbar \, D^j_{mm'}(\alpha\beta\gamma) \tag{8.33b}$$

$$K_z \, D^j_{mm'}(\alpha\beta\gamma) = m'\hbar \, D^j_{mm'}(\alpha\beta\gamma) \tag{8.33c}$$

$$H \, D^j_{mm'}(\alpha\beta\gamma) = E(j) \, D^j_{mm'}(\alpha\beta\gamma) \tag{8.33d}$$

where

$$E(j) = -\frac{\mu k^2}{2\hbar^2} \frac{1}{(2j+1)^2} \quad . \tag{8.33e}$$

The $D^j_{mm'}(\alpha\beta\gamma)$ are again the Wigner functions belonging to the irreducible representations of the rotation group. The angles (α,β,γ) describe the orientation of a rigid body in the earlier problem; in the next section it will be seen how they relate to the hydrogen atom. The quantum numbers (j,m,m') here refer to eigenvalues of the two commuting conserved vectors $\underset{\sim}{J}$ and $\underset{\sim}{K}$ which are derived from the conserved (but noncommuting) vectors $\underset{\sim}{A}$ and $\underset{\sim}{L}$. These are the three quantum numbers which correspond to (N,ℓ,m) in the usual treatment of the hydrogen atom. The latter represent the principal quantum number, the angular momentum, and one component of the angular momentum. In fact $N = 2j + 1$ as we have already seen.

Finally the degeneracy of this problem is exactly the same as that of the spherical top since the energy depends only on j, and not on m and m'. The degeneracy is therefore $(2j+1)^2 = N^2$. If one now takes into account the spin and the exclusion principle, one sees that the number of electrons in a closed shell is $2N^2$. These are the magic numbers of atomic physics that determine in zero approximation the number of elements in one period of the periodic table. The preceding formulation brings out the relation of the rotation group to the structure of the periodic system.

5.9 THE SCHRÖDINGER DESCRIPTION OF THE KEPLER PROBLEM

The Heisenberg equations just discussed make the algebraic structure very clear. However, we are left with $|D^j_{mm'}(\alpha\beta\gamma)|^2$ as the probability for the configuration (α,β,γ) and we have not yet learned what these angles represent in the Kepler problem. To discuss this question and also to connect with more familiar formulations of the Kepler problem, let us go over to the Schrödinger representation.

$$H \mid > = - \frac{\hbar}{i} \frac{\partial}{\partial t} \mid > \quad . \tag{9.1}$$

For steady states

$$H \mid > = E \mid > \quad . \tag{9.2}$$

This equation may be solved in the x-representation. Then the Schrödinger equation separates in the same way as the Hamilton-Jacobi equation, namely, in spherical and parabolic coordinates as the limiting cases of prolate spheroidal coordinates. One gets Laguerre functions and confluent hypergeometric functions in the spherical and parabolic cases respectively.

One may also make the following table[6]:

	Integrals of Motion	Separation	Degeneracy Classical	Degeneracy Quantum	Group
Central Potential	$\underset{\sim}{L}$	spherical	$\nu_1=\nu_2$	$2\ell + 1$	0_3
Coulomb Potential	$\underset{\sim}{L}$ and $\underset{\sim}{A}$	spherical parabolic (prolate spheroidal)	$\nu_1=\nu_2=\nu_3$	N^2	0_4

We shall now study equation (9.2) from the point of view of the symmetry group. It turns out to be most convenient to do this in the momentum representation. Let us start from

$$\left(E - \frac{p^2}{2\mu}\right) | \rangle = V | \rangle \quad . \tag{9.3}$$

Then

$$\langle \underset{\sim}{p} | \int \left(E - \frac{p^2}{2\mu}\right) |\underset{\sim}{p}'\rangle \langle \underset{\sim}{p}' | \rangle \, d\underset{\sim}{p}' = \langle \underset{\sim}{p} | \int V |\underset{\sim}{p}'\rangle \langle \underset{\sim}{p}' | \rangle \, d\underset{\sim}{p}'$$

and

$$\left(E - \frac{p^2}{2\mu}\right) \langle \underset{\sim}{p} | \rangle = \int \langle \underset{\sim}{p} | V | \underset{\sim}{p}'\rangle \langle \underset{\sim}{p}' | \rangle \, d\underset{\sim}{p}' \tag{9.4}$$

or

$$\left(E - \frac{p^2}{2\mu}\right) \phi(\underset{\sim}{p}) = \int \tilde{V}(\underset{\sim}{p}-\underset{\sim}{p}') \, \phi(\underset{\sim}{p}') \, d\underset{\sim}{p}' \tag{9.5}$$

where

$$\phi(\underset{\sim}{p}) = \langle \underset{\sim}{p} | \rangle \tag{9.6}$$

$$\tilde{V}(\underset{\sim}{p}-\underset{\sim}{p}') = \langle \underset{\sim}{p} | V | \underset{\sim}{p}' \rangle \quad . \tag{9.7}$$

Next define p_o by

$$\frac{p_o^{\,2}}{2\mu} = - E \quad . \tag{9.8}$$

We shall be interested only in the bound states so that $E < 0$ and p_o is real. Then

$$\left(1 + \frac{p^2}{p_o^2}\right) \phi(\underset{\sim}{p}) = \frac{1}{E} \int \tilde{V}(\underset{\sim}{p}-\underset{\sim}{p}') \phi(\underset{\sim}{p}') d\underset{\sim}{p}' \quad .$$

Define

$$G^{-1}(p) = 1 + \frac{p^2}{p_o^2} \quad . \tag{9.9}$$

Then

$$G^{-1}(p) \phi(\underset{\sim}{p}) = \frac{1}{E} \int \tilde{V}(\underset{\sim}{p}-\underset{\sim}{p}') \phi(\underset{\sim}{p}') d\underset{\sim}{p}' \tag{9.10}$$

and

$$G^{-2}(p) \phi(\underset{\sim}{p}) =$$

$$= \frac{1}{E} \int G^{-1}(p) \tilde{V}(\underset{\sim}{p}-\underset{\sim}{p}') G^{-1}(p') G^{-2}(p') \phi(\underset{\sim}{p}') G^{3}(p') d\underset{\sim}{p}'$$

or

$$\Phi(\underset{\sim}{p}) = \frac{1}{E} \int U(\underset{\sim}{p},\underset{\sim}{p}') \Phi(\underset{\sim}{p}') d\tau' \tag{9.11}$$

where

$$\Phi(\underset{\sim}{p}) = G^{-2}(p) \phi(\underset{\sim}{p}) \tag{9.12}$$

$$U(\underset{\sim}{p},\underset{\sim}{p}') = G^{-1}(p) \tilde{V}(\underset{\sim}{p}-\underset{\sim}{p}') G^{-1}(p') \tag{9.13}$$

$$d\tau = G^{3}(p) d\underset{\sim}{p} \quad . \tag{9.14}$$

Equation (9.11) is equivalent to (9.5) but the kernel is now symmetric: $U(p,p') = U(p',p)$. Here $\tilde{V}(p-p')$ is related to the original potential by a Fourier transformation as follows:

$$\tilde{V}(\underset{\sim}{p}-\underset{\sim}{p}') = \langle \underset{\sim}{p}|V|\underset{\sim}{p}'\rangle = \iint d\underset{\sim}{x}\, d\underset{\sim}{x}' \langle \underset{\sim}{p}|\underset{\sim}{x}\rangle \langle \underset{\sim}{x}|V|\underset{\sim}{x}'\rangle \langle \underset{\sim}{x}'|\underset{\sim}{p}'\rangle \quad . \tag{9.15}$$

The assumption of a local potential is that

$$\langle \underset{\sim}{x}|V|\underset{\sim}{x}'\rangle = V(\underset{\sim}{x})\, \delta(\underset{\sim}{x}-\underset{\sim}{x}') \tag{9.16}$$

so that

$$\begin{aligned} \langle \underset{\sim}{p}|V|\underset{\sim}{p}'\rangle &= \int d\underset{\sim}{x} \langle \underset{\sim}{p}|\underset{\sim}{x}\rangle V(\underset{\sim}{x}) \langle \underset{\sim}{x}|\underset{\sim}{p}'\rangle \\ &= (1/2\pi\hbar)^3 \int d\underset{\sim}{x} \exp\left[\frac{i}{\hbar}\, \underset{\sim}{x}(\underset{\sim}{p}'-\underset{\sim}{p})\right] V(\underset{\sim}{x}) \quad . \end{aligned} \tag{9.17}$$

We shall next make a group theoretical interpretation of (9.11).

5.10 RELATION TO THE ROTATION GROUP[7]

A rotation may be specified by a vector $\underset{\sim}{\omega}$ that is directed along the axis of rotation and that is equal in length to the magnitude of the rotation. Then all rotations may be represented by vectors from the origin to points in the interior of a sphere of radius π. This set of points is called the group space. Rotations may also be parameterized by Eulerian angles $(\alpha\beta\gamma)$ and in other ways. These different parameterizations may be regarded as different coordinate systems for the group space.

The Wigner functions form a complete orthogonal set over this space as follows[8]

$$\int_0^{2\pi}\int_0^{\pi}\int_0^{2\pi} D^j_{mn}(\alpha\beta\gamma)^* \, D^{j'}_{m'n'}(\alpha\beta\gamma)\ \rho(\alpha\beta\gamma)\ d\alpha\ d\beta\ d\gamma$$

$$= \delta^{jj'}\ \delta_{mm'}\ \delta_{nn'}\ \frac{c}{2j+1} \tag{10.1}$$

where the weight function is

$$\rho(\alpha\beta\gamma) = \sin\beta \tag{10.2}$$

and c is a constant.

If one changes to some other coordinate system (a) then the preceding equation becomes

$$\iiint \mathcal{D}^j_{mn}(a)^* \, \mathcal{D}^{j'}_{m'n'}(a)\ d\tau(a) = \delta^{jj'}\ \delta_{mm'}\ \delta_{nn'}\ \frac{c}{2j+1} \tag{10.3}$$

where

$$\mathcal{D}^j_{mn}(a) = D^j_{mn}(\alpha\beta\gamma)$$

$$d\tau(a) = \rho(\alpha\beta\gamma)\ d\alpha\ d\beta\ d\gamma = \rho(a)\ d\underset{\sim}{a} \quad . \tag{10.4}$$

Here $d\tau(a)$ is called the invariant volume element. By choosing $j = j' = 0$ we see that

$$\int d\tau = c \quad .$$

Therefore c is the total volume of the group space.

Now define

$$K(a,a') = \frac{1}{c} \sum_{jmn} \mathcal{D}^j_{mn}(a)\, \mathcal{D}^j_{mn}(a')^* \quad . \tag{10.5}$$

Then

$$\int K(a,a')\, \mathcal{D}^{j'}_{m'n'}(a')\, d\tau(a')$$

$$= \sum \frac{1}{c} \mathcal{D}^j_{mn}(a) \int \mathcal{D}^j_{mn}(a')^* \, \mathcal{D}^{j'}_{m'n'}(a')\, d\tau(a')$$

$$= \frac{1}{c} \mathcal{D}^{j'}_{m'n'}(a) \left(\frac{c}{2j+1}\right)$$

or

$$\int K(a,a')\, \mathcal{D}^j_{mn}(a')\, d\tau' = \frac{1}{2j+1} \mathcal{D}^j_{mn}(a) \quad . \tag{10.6}$$

Therefore $\mathcal{D}^j_{mn}(a)$ is an eigenfunction of this integral equation. We now want to show that equations (9.11) and (10.6) are in fact equivalent when the potential is Coulombic. This remark obviously depends on a particular correspondence between momentum space and group space and is the following:

$$U = D^{1/2}(\alpha\beta\gamma) = \frac{p_o + i\underset{\sim}{p}\underset{\sim}{\sigma}}{p_o - i\underset{\sim}{p}\underset{\sim}{\sigma}} \tag{10.7}$$

where U is the 2 × 2 unitary representation of the rotation group. This equation then establishes a correspondence between a point in group space (α,β,γ) and a point in momentum space $(\underset{\sim}{p}/p_o)$. The possibility of making this correspondence depends just on the possibility of representing a unitary matrix in the form

$$U = \frac{1 + iH}{1 - iH}$$

where H is hermitian.

Using the coordinate system $(\underset{\sim}{p}/p_o)$ introduced in (10.7), one may now show that the invariant volume element (10.4) is

$$d\tau(a) = G^3(p)\ d\underset{\sim}{p} \tag{10.8}$$

where G(p) is just the function given earlier in (9.9).[9] Therefore equation (9.11) may be regarded as an integral equation on the group space. The three coordinates of a point in this space are now $\underset{\sim}{p}/p_o$ and the integration over the complete group space is an integration over all of momentum space. In this way the Schrödinger eigenvalue problem may be expressed as an eigenvalue problem on the group space.

So far no assumption has been made about the nature of the potential. However, if the potential is Coulombic, then its Fourier transform is simply related to the kernel of (10.6). In fact one may show[10]

$$U(p,p') = b\, p_o\, K(p,p') , \qquad b = -e^2/2\hbar \tag{10.9}$$

where $U(p,p')$ and $K(p,p')$ are defined by (9.13) and (10.5) respectively.

If one then substitutes (10.9) into (9.11) one obtains

$$\Phi(p) = \frac{1}{E} \int b\, p_o\, K(p,p')\, \Phi(p')\, d\tau' \tag{10.10}$$

which is just the same as (10.6).

Therefore

$$\frac{bp_o}{E} = 2j + 1 \tag{10.11}$$

and

$$\Phi(p) \sim D^j_{mn}(p) \quad . \tag{10.12}$$

From (10.11) one finds

$$b\, p_o = (2j + 1) \left(- \frac{p_o{}^2}{2\mu}\right)$$

or

$$p_o = -\frac{2\mu b}{2j+1}$$

and

$$E = -\frac{p_o^2}{2\mu}$$

$$= -\frac{2\mu b^2}{(2j+1)^2} = -\frac{1}{2}\mu c^2\left(\frac{e^2}{\hbar c}\right)^2\frac{1}{(2j+1)^2} \qquad (10.13)$$

which is again the Balmer formula.

The wave functions are given by (10.12)

$$\phi^j_{mn}(\underset{\sim}{p}) \sim G^2(p)\,\mathcal{D}^j_{mn}(\underset{\sim}{p}) \quad . \qquad (10.14)$$

In (10.14) we have disposed of the question left unanswered by the algebraic analysis, namely: what is the interpretation of $G^2(p)\ D^j_{mn}(\alpha\beta\gamma)$? According to (10.14) these are probability amplitudes in momentum space when $(\alpha\beta\gamma)$ are related to $\underset{\sim}{p}/p_o$ by (10.7). It should be noted again that ϕ^j_{mn} are eigenfunctions of J^2, J_z, K_z, but not of the angular momentum. These are in fact the Fourier transforms of the functions that arise when the Schrödinger equation is separated in parabolic coordinates.

To obtain their relation to the functions which arise

when the separation is made in spherical coordinates, express the most general wavefunction which can be represented in the space of these functions with a given energy.

$$| \,\rangle = \sum_{mn} |jmn\rangle \langle jmn| \,\rangle \quad . \tag{10.15}$$

There is no sum over the energy index (j).

This space includes eigenfunction of angular momentum $|LM\rangle$. Therefore

$$|NLM\rangle = \sum_{mn} |jmn\rangle \langle jmn|NLM\rangle \tag{10.16}$$

or

$$\Pi_{NL}(p)\ Y_{LM}(\theta,\phi) = \sum_{mn} c(jmn,\ NLM)\ \phi^{j}_{mn}(\underset{\sim}{p}) \tag{10.17}$$

where

$$\langle p\theta\phi|NLM\rangle = \Pi_{NL}(p)\ Y_{LM}(\theta,\phi) \tag{10.18}$$

and

$$\phi^{j}_{mn}(\underset{\sim}{p}) = \langle \underset{\sim}{p}|jmn\rangle \quad . \tag{10.19}$$

Every function in this space satisfies the integral equation (9.11). Therefore

$$\Pi_{NL}(p)\ Y_{LM}(\theta,\phi) = \frac{1}{E}\int U(p,p')\ \Pi_{NL}(p')\ Y_{LM}(\theta',\phi')\ d\tau' \quad .$$

By carrying out the angular integration, one finds

$$\Pi_{NL}(p) = \frac{1}{E}\int k(p,p')\ \Pi_{NL}(p')\ d\tau'_p \tag{10.20}$$

where

$$d\tau_p = [g(p)]^{1/2}\ p^2\ dp \tag{10.21}$$

and

$$k(p,p') = \frac{e^2}{\pi\hbar}\,\frac{1}{pp'}\,\frac{1}{G(p)\ G(p')}\ Q_L\left(\frac{p^2 + p'^2}{2pp'}\right) \quad . \tag{10.22}$$

The Q_L are associated Legendre functions of the second kind, and the $\Pi_{NL}(p)$ are Gegenbauer functions.[11] The Gegenbauer functions are Fourier transforms of the Laguerre functions which arise when the Schrödinger equation is solved in spherical coordinates. The $G^2(p)\ \mathcal{D}^j_{mn}(\alpha\beta\gamma)$, on the other hand, are the Fourier transforms of the confluent hypergeometric functions that arise when the Schrödinger equation is solved in parabolic coordinates.

5.11 CLASSICAL TREATMENT OF SCATTERING[12]

If the energy is positive, one finds that the classical Kepler orbit is hyperbolic according to the classification following (3.26). For any other attractive or repulsive potential the paths still begin and end at infinity provided that the energy is still positive.

The orbit in a general central potential is, according to (4.20) of this chapter, given by

$$\theta = L \int_r^\infty \left[2\mu(E-V) - \frac{L^2}{r^2} \right]^{-1/2} \frac{dr}{r^2} \tag{11.1}$$

where L and E are the angular momentum and the energy, and where θ and r are shown in the figure. Let θ_A and r_A be the polar coordinates of A, the point of closest approach.

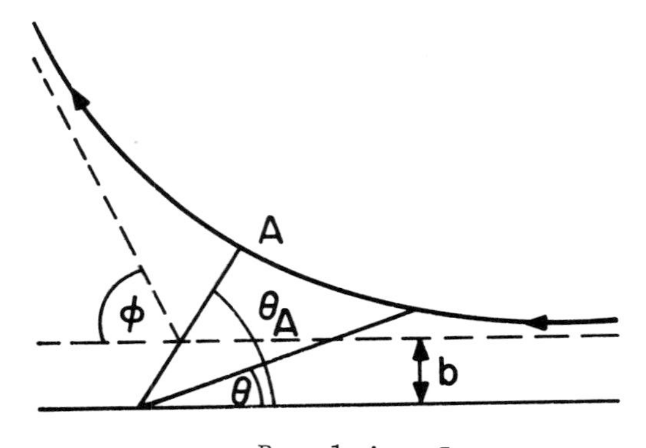

Repulsive Center

Fig. 11.1. Classical Orbit

Let us also introduce as the initial data p_∞, the momentum at ∞, and b, the impact parameter, the distance at which the particle would pass if there were no potential. Then

$$2\mu E = p_\infty^2 \tag{11.2}$$

$$L = b\, p_\infty \quad . \tag{11.3}$$

Also let ϕ be the total deflection angle:

$$\phi = \pi - 2\theta_A \quad . \tag{11.4}$$

Substitute for E, L, and ϕ in (11.1):

$$\frac{1}{2}(\pi-\phi) = (bp_\infty)\int_{r_A}^{\infty} [p_\infty^2 - 2\mu V - b^2 p_\infty^2/r^2]^{-1/2}\frac{dr}{r^2} \quad . \tag{11.5}$$

Then the preceding equation determines the deflection angle, ϕ, as a function of the impact parameter, b.

We are interested in the scattering not of a single particle but of a beam of identical particles. Assume that the different particles in the beam have the same energy but different impact parameters so that they are scattered through different angles ϕ.

Let n be the particle current in the incident beam (number of particles per unit area per unit time). Then the total flux through a ring (b, b+db) perpendicular to the incident beam is

$$dN = (2\pi\ b\ db)\ n \qquad . \tag{11.6}$$

The particles passing through the ring are then scattered through angles between $\left(\phi(b),\ \phi(b) + d\phi(b)\right)$ as determined by (11.5). Now define the cross section $d\sigma$ to be the number of particles scattered into the angular interval $(\phi,\ \phi+d\phi)$ divided by the incident current:

$$d\sigma = dN/n = 2\pi\ b\ db$$

by (11.6). Or

$$\frac{d\sigma}{d\phi} = 2\pi b\ \frac{db}{d\phi} \qquad . \tag{11.7}$$

The differential cross section is usually referred to solid angle. Let $d\omega$ be the solid angle bounded by the cones with vertical angles ϕ and $\phi + d\phi$. Then

$$d\omega = 2\pi\ \sin\phi\ d\phi \tag{11.8}$$

and therefore (11.7) becomes

$$\frac{d\sigma}{d\omega} = \frac{b}{\sin\phi}\ \frac{db}{d\phi} \qquad . \tag{11.9}$$

Equation (11.9) is the classical formula for the differential cross section. With the aid of (11.5) one may compute $db/d\phi$ and therefore the differential cross section for any given potential $V(r)$.

As an example let us consider the Newtonian or the Coulomb potential. We may easily integrate (11.5) for $V(r) = k/r$. It is, however, again more interesting to consider the Runge-Lenz vector

$$\mathbf{A} = \mathbf{L} \times \mathbf{p} + \mu k\, \hat{\mathbf{r}} \quad .$$

We found the orbits in paragraph 3 by looking at the scalar product $\mathbf{A}\mathbf{r}$. Since we are now interested in the momentum, it is natural to examine the projection

$$\mathbf{A}\mathbf{p} = \mu k\, \hat{\mathbf{r}}\, \mathbf{p} \quad . \tag{11.10}$$

Here $\mathbf{A}$ is a constant vector (directed along the major axis) and equation (11.10) holds for all $(\hat{r},\mathbf{p})$. In particular

$$\mathbf{A}\, \mathbf{p}_\infty = \mu k\, \hat{\mathbf{r}}_\infty\, \mathbf{p}_\infty \tag{11.11}$$

or

$$A\, p_\infty \cos \theta_A = \mu k\, p_\infty$$

since

$$\hat{\mathbf{r}}_\infty \,||\, - \mathbf{p}_\infty \quad . \tag{11.12}$$

Therefore

$$\cos \theta_A = \mu k/A \ (= 1/\varepsilon) \tag{11.13}$$

where θ_A is the angle that $-\underset{\sim}{p}_\infty$ makes with $\underset{\sim}{A}$ and therefore with the major axis of the hyperbola. [The eccentricity, ε, appearing in (11.13) is given by (3.22). In terms of the eccentricity, (11.13) is geometrically obvious.]

By (3.25)

$$1 = \left(\frac{\mu k}{A}\right)^2 + \frac{2\mu L^2 E}{A^2} \quad . \tag{11.14}$$

By (11.13) and (11.14)

$$\sin \theta_A = \frac{L(2\mu E)^{1/2}}{A} = \frac{Lp_\infty}{A}$$

and

$$\tan \theta_A = \frac{Lp_\infty}{\mu k} \quad . \tag{11.15}$$

But $\phi = \pi - 2\theta_A$ and therefore by the preceding equation

$$\cot \frac{\phi}{2} = \frac{Lp_\infty}{\mu k} = \frac{bp_\infty{}^2}{\mu k} \quad . \tag{11.16}$$

It now follows from the general formula for the differential cross section (11.9) that

$$\frac{d\sigma}{d\omega} = \left(\frac{k}{4E}\right)^2 \frac{1}{\sin^4 \frac{1}{2}\phi} \quad . \tag{11.17}$$

This is the famous Rutherford formula, valid for the gravitational scattering of stars as well as the Coulomb scattering of electrons.

5.12 QUANTUM TREATMENT OF SCATTERING[13,14]

The corresponding quantum problem is of much greater practical importance than the classical limit, since experimental scattering techniques are so important in both atomic and nuclear physics. Although the systematic discussion of such questions is beyond the scope of this text, we should like to conclude with a few remarks about the formulation of the quantum problem.

One requires a solution of the following equation

$$(\overset{\circ}{H} + V)|\rangle = E|\rangle \tag{12.1a}$$

or

$$(E - \overset{\circ}{H})|\rangle = V|\rangle \tag{12.1b}$$

when the energy, E, is positive. Here V is the scattering

potential, $\overset{\circ}{H}$ is the energy of the free particle:

$$\overset{\circ}{H} = \frac{1}{2\mu}\left(\frac{\hbar}{i}\underset{\sim}{\nabla}\right)^2 \tag{12.2}$$

and μ is the reduced mass. The problem may be formulated in terms of the integral equation:

$$|\ \rangle = |k\rangle + GV|\ \rangle \tag{12.3}$$

where G is a Green's function:

$$G^{-1} = E - \overset{\circ}{H} \tag{12.4}$$

and $|k\rangle$ represents an incident plane wave so that

$$(E - \overset{\circ}{H})|k\rangle = 0 \tag{12.5}$$

and consequently

$$(E - \overset{\circ}{H})|\ \rangle = (E - \overset{\circ}{H})|k\rangle + V|\ \rangle = V|\ \rangle$$

by (12.3) and the definition of $|k\rangle$ and G. Therefore (12.3) is equivalent to the differential equation (12.1) but in addition contains the boundary condition

$$|\ \rangle = |k\rangle$$

to specify the initial wave function.

In the momentum representation the Green's function is

$$\langle \underset{\sim}{k}'|G|\underset{\sim}{k}''\rangle = \langle \underset{\sim}{k}'| \frac{1}{E-\overset{\circ}{H}} |\underset{\sim}{k}''\rangle = \frac{2\mu}{\hbar^2} \frac{1}{\underset{\sim}{k}^2-\underset{\sim}{k}'^2} \delta(\underset{\sim}{k}-\underset{\sim}{k}') \qquad (12.6)$$

and in the x-representation one finds

$$\langle \underset{\sim}{x}'|G|\underset{\sim}{x}''\rangle = \iint \langle \underset{\sim}{x}'|\underset{\sim}{k}'\rangle \, d\underset{\sim}{k}' \, \langle \underset{\sim}{k}'|G|\underset{\sim}{k}''\rangle \, d\underset{\sim}{k}'' \, \langle \underset{\sim}{k}''|\underset{\sim}{x}''\rangle$$

$$= \frac{2\mu}{\hbar^2} \left(\frac{1}{2\pi}\right)^3 \int e^{i\underset{\sim}{k}'(\underset{\sim}{x}'-\underset{\sim}{x}'')} \frac{1}{\underset{\sim}{k}^2-\underset{\sim}{k}'^2} \, d\underset{\sim}{k}' \; . \qquad (12.7)$$

The indicated integration runs over the point k = k' and therefore the above Green's function is so far not completely defined. Here we wish to describe outgoing scattered waves and shall complete the definition by specifying the path of integration as follows:

$$G_+ = \frac{1}{E - \overset{\circ}{H} + i\varepsilon} \; . \qquad (12.8)$$

Then

$$\langle \underset{\sim}{x}'|G_+|\underset{\sim}{x}''\rangle = \frac{2\mu}{\hbar^2} \left(\frac{1}{2\pi}\right)^3 \int e^{i\underset{\sim}{k}'(\underset{\sim}{x}'-\underset{\sim}{x}'')} \frac{1}{\underset{\sim}{k}^2 - \underset{\sim}{k}'^2 + i\varepsilon} \, d\underset{\sim}{k}'$$

$$= -\frac{2\mu}{\hbar^2} \frac{1}{4\pi} \frac{e^{ik|\underset{\sim}{x}'-\underset{\sim}{x}''|}}{|\underset{\sim}{x}'-\underset{\sim}{x}''|} \; . \qquad (12.9)$$

The integral equation (12.3) may now be expressed in the k-representation:

$$\langle \underset{\sim}{k}'| \;\rangle = \langle \underset{\sim}{k}'|\underset{\sim}{k}\rangle + \langle \underset{\sim}{k}'|GV| \;\rangle$$
$$= \langle \underset{\sim}{k}'|\underset{\sim}{k}\rangle + \int \langle \underset{\sim}{k}'|G|\underset{\sim}{k}''\rangle \, d\underset{\sim}{k}'' \, \langle \underset{\sim}{k}''|V|\underset{\sim}{k}'''\rangle \, d\underset{\sim}{k}''' \, \langle \underset{\sim}{k}'''| \;\rangle$$

or

$$\phi(\underset{\sim}{k}') = \delta(\underset{\sim}{k}'-\underset{\sim}{k}) + \frac{2\mu}{\hbar^2} \int \frac{1}{\underset{\sim}{k}^2 - \underset{\sim}{k}'^2 + i\varepsilon} \langle \underset{\sim}{k}'|V|\underset{\sim}{k}''\rangle \, d\underset{\sim}{k}'' \, \phi(\underset{\sim}{k}'') \tag{12.10}$$

where

$$\phi(\underset{\sim}{k}') = \langle \underset{\sim}{k}'| \;\rangle \quad .$$

$\phi(\underset{\sim}{k}')$ is the probability amplitude in momentum space and therefore $|\phi(\underset{\sim}{k}')|^2$ is the probability of $\underset{\sim}{k}'$.

In the x-representation (12.3) becomes

$$\langle \underset{\sim}{x}'| \;\rangle = \langle \underset{\sim}{x}'|\underset{\sim}{k}\rangle + \int \langle \underset{\sim}{x}'|G|\underset{\sim}{x}''\rangle \, d\underset{\sim}{x}'' \, \langle \underset{\sim}{x}''|V|\underset{\sim}{x}'''\rangle \, d\underset{\sim}{x}''' \, \langle \underset{\sim}{x}'''| \;\rangle$$

or by (12.9)

$$\psi(\underset{\sim}{x}') = \psi_k(\underset{\sim}{x}') - \frac{\mu}{2\pi\hbar^2} \int \frac{e^{ik|\underset{\sim}{x}'-\underset{\sim}{x}''|}}{|\underset{\sim}{x}'-\underset{\sim}{x}''|} V(\underset{\sim}{x}'') \, \psi(\underset{\sim}{x}'') \, d\underset{\sim}{x}'' \tag{12.11}$$

where

$$\psi(\underset{\sim}{x}') = \langle \underset{\sim}{x}'| \;\rangle \quad , \qquad \psi_k(\underset{\sim}{x}') = \langle \underset{\sim}{x}'|\underset{\sim}{k}\rangle \quad .$$

The asymptotic form of (12.11), when $\underset{\sim}{x}'$ is far from the scattering center, is

$$\psi(x') \to \psi_k(\underset{\sim}{R}) + \frac{e^{ikR}}{R} f(\theta,\phi) \tag{12.12}$$

where

$$f(\theta,\phi) = -\frac{\mu}{2\pi\hbar^2} \int e^{-i\underset{\sim}{k}'\underset{\sim}{x}} V(\underset{\sim}{x})\ \psi(\underset{\sim}{x})\ d\underset{\sim}{x} \quad . \tag{12.13}$$

Here large $\underset{\sim}{x}'$ is replaced by $\underset{\sim}{R}$ and $\underset{\sim}{k}' \equiv k\hat{R}$.

We may then write

$$\psi = \psi_{\underset{\sim}{k}} + \psi_{sc} \tag{12.14}$$

where $\psi_{\underset{\sim}{k}}$ represents the incident plane wave and ψ_{sc} represents the scattered wave. Then the incident current is parallel to $\underset{\sim}{k}$ and equal to

$$\underset{\sim}{S}_{inc} = \frac{1}{\mu} \text{Re}\ \psi^*_k\ \underset{\sim}{p}\ \psi_k \tag{12.15}$$

and the scattered current is radial and equal to

$$\underset{\sim}{S}_{sc} = \frac{1}{\mu} \text{Re}\ \psi^*_{sc}\ \underset{\sim}{p}\ \psi_{sc} \quad . \tag{12.16}$$

The corresponding differential cross section may be computed from (12.12) since it is by definition the ratio of the scattered to the incident currents. Then

$$\frac{d\sigma}{d\omega} = |f(\theta,\phi)|^2 \quad . \tag{12.17}$$

By (12.13)

$$f(\theta,\phi) = -\frac{\mu}{2\pi\hbar^2} \langle \underset{\sim}{k}'|V| \;\rangle \quad . \tag{12.18}$$

Define the transition matrix, T, by

$$\langle \underset{\sim}{k}''|T|\underset{\sim}{k}'\rangle = \langle \underset{\sim}{k}''|V| \;\rangle \tag{12.19}$$

where

$$| \;\rangle = |\underset{\sim}{k}'\rangle + GV| \;\rangle \quad . \tag{12.20}$$

Then

$$f(\theta,\phi) = -\frac{\mu}{2\pi\hbar^2} \langle \underset{\sim}{k}''|T|\underset{\sim}{k}'\rangle \tag{12.21}$$

and

$$\frac{d\sigma}{d\omega} = \left(\frac{\mu}{2\pi\hbar^2}\right)^2 |\langle \underset{\sim}{k}''|T|\underset{\sim}{k}'\rangle|^2 \quad . \tag{12.22}$$

This last formula is an exact expression for the differential cross section in terms of the above T-matrix or in terms of the exact solution $| \;\rangle$ of the integral equation (12.20). From (12.20) one may also obtain an integral equation for T itself as follows:

$$\langle \underset{\sim}{k}''|T|\underset{\sim}{k}'\rangle = \langle \underset{\sim}{k}''|V|\ \rangle = \langle \underset{\sim}{k}''|V|\underset{\sim}{k}'\rangle + \langle \underset{\sim}{k}''|VGV|\ \rangle$$

or

$$T = V + VGT \quad . \tag{12.23}$$

By iteration one obtains the Born series

$$T = V + VGV + VGVGV + \cdots \quad . \tag{12.23a}$$

Written out as an integral equation (12.23) becomes

$$\langle \underset{\sim}{k}''|T|\underset{\sim}{k}'\rangle = \langle \underset{\sim}{k}''|V|\underset{\sim}{k}'\rangle + \frac{2\mu}{\hbar^2}\int \langle \underset{\sim}{k}''|V|\underset{\sim}{k}\rangle \frac{1}{\underset{\sim}{k}'-\underset{\sim}{k}^2+i\varepsilon} \times \langle \underset{\sim}{k}|T|\underset{\sim}{k}'\rangle\, d\underset{\sim}{k} \quad . \tag{12.23b}$$

Then (12.22) and (12.23) together indicate a procedure for computing the differential cross section. Equation (12.23b) is also known as the Lippmann-Schwinger equation.

The first approximation to T is simply V itself and the corresponding differential cross section is

$$\frac{d\sigma}{d\omega} = \left(\frac{\mu}{2\pi\hbar^2}\right)^2 |\langle \underset{\sim}{k}'|V|\underset{\sim}{k}\rangle|^2 \quad . \tag{12.24}$$

This is the first Born approximation. It is valid when one can put $|\ \rangle = |\underset{\sim}{k}'\rangle$ in (12.19), or in other words, when the potential distorts the incident plane wave only slightly.

If one substitutes in (12.24) the screened Coulomb potential, and then lets the screening go to zero, one recovers the classical Rutherford formula (12.17).

In the classical limit the wave packet must move in the classical orbit. One may obtain such an approximation to the classical motion by making the ansatz

$$\psi(x) \sim e^{iS(x)/\hbar} \tag{12.25}$$

in (12.13) or (12.19) as we did earlier in Chapter 3 [paragraph (3.7)]. Then

$$\langle \underset{\sim}{k}'|T|\underset{\sim}{k}\rangle = \int e^{-i\underset{\sim}{k}'\underset{\sim}{x}} V(\underset{\sim}{x})\, e^{iS/\hbar}\, d\underset{\sim}{x} \tag{12.26}$$

where S(x) is the classical action and satisfies the Hamilton-Jacobi equation

$$\frac{1}{2\mu}(\underset{\sim}{\nabla}S)^2 + V(x) = E \quad . \tag{12.27}$$

The classical orbits are orthogonal to the surfaces of constant S as discussed in Chapter 3, section 6C. The exact orbits are here given by (11.1). When the energy is very high, however, the incident particle tends to be only slightly deviated by the potential and one may try to approximate S by assuming straight line classical motion. By following this argument one may obtain the so-called eikonal approximation in the following form[14]:

$$f(\theta) = -\, ik \int_0^\infty [e^{2i\Delta(b)} - 1]\, J_o(kb)\, b\, db \tag{12.28}$$

where

$$\Delta(b) = -\frac{1}{4} \int_{-\infty}^{\infty} \frac{V\left((b^2+z^2)^{1/2}\right)}{E}\, d(kz) \quad . \tag{12.29}$$

In these formulas b is the impact parameter introduced in the preceding classical treatment. To obtain the phase $\Delta(b)$ integrate from $z = -\infty$ to $z = +\infty$ along a straight line path with impact parameter b. The total scattering amplitude is then represented as a sum of coherent contributions from these different classical paths.

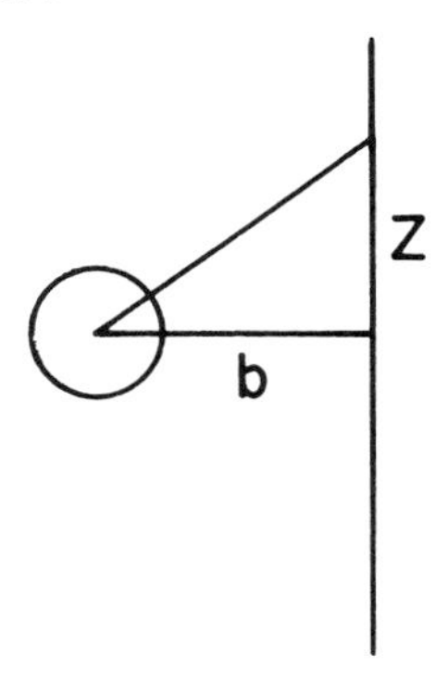

Fig. 12.1. Single Path Contribution to Amplitude (12.28).

NOTES ON CHAPTER 5

1. The central field approximation is very accurate in the planetary case and consequently the actual orbits may be treated as slowly precessing ellipses. For example, in the case of the planet Mercury, the rate of the advance of the perihelion is calculated to be 529.2 seconds of arc per century and is mainly due to perturbations by Venus (276 sec.), Jupiter (153 sec.), and Earth (90 sec.). The observed precession of Mercury is somewhat larger (572.7 sec. per century) but the discrepancy between 572.7 and 529.2 can be understood as a correction required by general relativity.

   The central field approximation is also a very satisfactory way to begin the analysis of the electronic structure of atoms. In the next stage of the perturbation method one takes the interaction of electrons into account by the Hartree-Fock method.

2. Hamilton's principal function, S, is related to W, Hamilton's characteristic function, by $S = W - \alpha_1 t$ as in (6.6) of Chapter 3.

3. See Dirac, op. cit.

   We shall briefly describe how the action and angle variables are related to the emission and absorption operators.

The action J is a momentum and the angle ω is conjugate to it. Classically,

$$[\omega, J] = 1 \quad .$$

In quantum theory one defines for the harmonic oscillator

$$e^{i\omega} = \begin{matrix} 0 & 0 & 0 & \\ 1 & 0 & 0 & \cdots \\ 0 & 1 & 0 & \end{matrix} \qquad e^{-i\omega} = \begin{matrix} 0 & 1 & 0 & 0 & \\ 0 & 0 & 1 & 0 & \cdots \\ 0 & 0 & 0 & 1 & \end{matrix}$$

$$J = \frac{H}{\omega} - \frac{1}{2}\hbar = \begin{matrix} 0 & 0 & 0 \\ 0 & 1 & 0 \\ 0 & 0 & 2 \end{matrix} \quad .$$

Then

$$(e^{i\omega}, J) = -\hbar\, e^{i\omega}$$

$$(e^{-i\omega}, J) = \hbar\, e^{-i\omega} \quad .$$

The classical limits of these formulas are

$$[e^{i\omega}, J] = i\, e^{i\omega}$$

$$[e^{-i\omega}, J] = -i\, e^{-i\omega}$$

where J and ω are the classical action and angle variables.

The absorption and emission operators may then be defined to be

$$a = J^{1/2}\, e^{i\omega} , \qquad \bar{a} = e^{-i\omega}\, J^{1/2}$$

with the properties

$$a\bar{a} = J , \qquad \bar{a}a = J + \hbar , \qquad (\bar{a}, a) = \hbar \quad .$$

In many body theory J plays the role of the number operator.

4. The action-angle variables are known in celestial mechanics as the Delaunay elements of an orbit. They specify the orientation, size, and shape of a planetary orbit and vary slowly in time in the presence of small perturbations.

   The same method of action-angle variables was taken over by the old quantum theory in the Sommerfeld-Wilson quantization conditions:

$$J_k = n_k\, h \quad .$$

The J_k are also the adiabatic-invariants of Ehrenfest. See also problem 21, Chapter 3.

   For a more detailed discussion of the Delaunay method see Goldstein and particularly Lanczos, op. cit. For adiabatic invariants see Landau and Lifshitz, op. cit. (Classical Mechanics, page 154).

5. P. Stehle and M. Han, Phys. Rev. 159, 1076 (1967).

6. The last column of this table refers to the symmetry group. The general central potential has the symmetry group O_3, generated by the three components of $\underset{\sim}{L}$; while the Coulomb potential has the symmetry group O_4, generated by the three components of $\underset{\sim}{A}$ as well as the three components of $\underset{\sim}{L}$.
7. This section depends on certain relations that are proved in Appendix E and further discussed in R. Finkelstein, J. Math. Phys. 8, 443 (1967).
8. See Edmonds, op. cit., for example.
9. See Appendix E, equation (E.47).
10. See Appendix E, equation (E.58) and problem 13 at end of this chapter.
11. See reference in note 7, equation (7.13).
12. See, for example, Landau and Lifshitz, op. cit.
13. See, for example, Saxon, also Gottfried, op. cit.
14. Gottfried, page 115.

PROBLEMS ON CHAPTER 5.

1. The transformation to the center of mass and relative coordinates is

$$M \underset{\sim}{X} = m_1 \underset{\sim}{x}_1 + m_2 \underset{\sim}{x}_2 , \qquad \underset{\sim}{x} = \underset{\sim}{x}_1 - \underset{\sim}{x}_2 \quad .$$

Show that the corresponding transformation on the conjugate variables is

$$\underset{\sim}{P} = \underset{\sim}{p}_1 + \underset{\sim}{p}_2 , \qquad \mu^{-1} \underset{\sim}{p} = m_1^{-1} \underset{\sim}{p}_1 - m_2^{-1} \underset{\sim}{p}_2$$

and show that

$$\frac{1}{2m_2} \underset{\sim}{p}_1^{\,2} + \frac{1}{2m_2} \underset{\sim}{p}_2^{\,2} = \frac{1}{2M} \underset{\sim}{P}^2 + \frac{1}{2\mu} \underset{\sim}{p}^2$$

$$\underset{\sim}{x}_1 \times \underset{\sim}{p}_1 + \underset{\sim}{x}_2 \times \underset{\sim}{p}_2 = \underset{\sim}{X} \times \underset{\sim}{P} + \underset{\sim}{x} \times \underset{\sim}{p} \quad .$$

2. Show that (3.21) describes a conic with eccentricity ε. Find the major axis of the ellipse.

3. Let the central potential be

$$V = -\frac{k}{r} + \phi(r)$$

where $\phi(r)$ is a small perturbation. Find $d\underset{\sim}{A}/dt$ and the rate at which the ellipse precesses.

4. Show that α_θ defined by (4.11) is the angular momentum.

5. Carry out the integrations indicated in (6.4) in general, and in (6.5) when $V(r)$ is a Newtonian potential.

6. Calculate the Poisson bracket (8.9) with the aid of (8.10).

7. Find the frequencies of a three dimensional harmonic oscillator with unequal force constants by the method of action-angle variables.

8. In the Kepler problem determine the action-angle variables corresponding to the independent frequencies by the method of Section 7.

   Do the corresponding work for the preceding problem in case that the three force constants become equal. (Use spherical polar coordinates.)

9. Integrate (4.20) to obtain the orbit in the Kepler problem. Express α_1 and α_θ in terms of J_2' and J_3'. Show that the semi-major axis, a, and the eccentricity ε are given by

$$a = \frac{J_3'^2}{4\pi^2 mk} , \qquad \varepsilon = \left[1 - \left(\frac{J_2'}{J_3'} \right)^2 \right]^{1/2} .$$

10. The motion of a degenerate harmonic oscillator can be separated in any Cartesian coordinate system. Obtain the transformation induced on action-angle variables, and corresponding emission and absorption variables, by a rotation of the Cartesian coordinate system.

11. A particle of charge e moves in an attractive central potential V(r) and a uniform magnetic field $\underset{\sim}{B}$. Show that the Hamiltonian is

$$H = \frac{\underset{\sim}{p}^2}{2m} + V(r) - \underset{\sim}{\mu}\underset{\sim}{B}$$

if the field is weak and μ is the magnetic moment.

Let $V(r)$ be a Coulomb potential so that the three frequencies generally present are equal if $\underset{\sim}{B} = 0$. Solve the problem for the case $\underset{\sim}{B} \neq 0$ by the action-angle method and compute the three frequencies in this case.

12. Prove (10.8), namely:

$$\sin\beta \, d\alpha \, d\beta \, d\gamma = \left(\frac{p_o^2}{p_o^2 + p^2} \right)^3 d\underset{\sim}{p}$$

where (α, β, γ) are related to $(\underset{\sim}{p}_o, \underset{\sim}{p})$ by (10.7).

13. Verify the following relation, implied by (10.9) of text:

$$G^{-1}(p) \, \tilde{V}(p-p') \, G(p') = \frac{b}{V_o} p_o \sum \mathcal{D}^j_{mn}(p) \, \mathcal{D}^j_{mn}(p')^*$$

where V_o is the volume of the group space

$$\tilde{V}(p-p') = \frac{1}{(2\pi h)^3} \int d\underset{\sim}{x} \, e^{(i/\hbar)(\underset{\sim}{p}'-\underset{\sim}{p})\underset{\sim}{x}} \, V(\underset{\sim}{x}) \quad .$$

Here $V(x) = -e/r$ and $b = -e^2/2\hbar$. See (E.58) of Appendix.

14. Prove (10.20) and show that $\Pi_{NL}(p)$ satisfies differential equation for Gegenbauer functions.

15. Since $\underset{\sim}{A}$ remains constant during Coulomb scattering, the initial and final momenta are connected with the corresponding angles by the following equation

$$\underset{\sim}{L}(\underset{\sim}{p}_{+\infty} - \underset{\sim}{p}_{-\infty}) + \mu k\ (\hat{\underset{\sim}{r}}_{+\infty} - \hat{\underset{\sim}{r}}_{-\infty}) = 0$$

where $(-\infty)$ and $(+\infty)$ refer to the incoming and the outgoing particles at ∞. From this relation obtain equation (11.16) of text.

16. Obtain the Rutherford formula by assuming a screened Coulomb potential

$$V = \frac{k\ e^{-r/a}}{r}$$

and making use of the Born formula (12.24).

17. Express the Lippmann-Schwinger equation as an integral equation on the group space by following the procedure of paragraph (5.10). R. Finkelstein and D. Levy, J. Math. Phys. 8, 2147 (1967).

APPENDIX A

HERMITIAN AND UNITARY MATRICES

We collect here some simple facts about hermitian and unitary matrices.

Definition:

The matrices A and B are <u>adjoint</u> to each other if

$$A_{rs} = B^*_{sr} \tag{A.1}$$

where * means complex conjugate.

The adjoint will be denoted by a dagger:

$$A^\dagger = B \tag{A.1a}$$

$$A = B^\dagger \quad . \tag{A.1b}$$

Remarks:

$$(A+B)^\dagger = A^\dagger + B^\dagger \tag{A.2}$$

$$(AB)^{\dagger} = B^{\dagger}A^{\dagger} \quad . \tag{A.3}$$

If A is diagonal,

$$A^{\dagger} = A^{*} \quad . \tag{A.4}$$

Definition:

A is <u>hermitian</u> if

$$A = A^{\dagger} \quad . \tag{A.5}$$

Definition:

A is unitary if

$$A^{-1} = A^{\dagger} \quad . \tag{A.6}$$

Remarks:

If H_1 and H_2 are hermitian, then $H_1 + H_2$ is hermitian:

$$(H_1+H_2)^{\dagger} = H_1^{\dagger} + H_2^{\dagger} = H_1 + H_2 \quad . \tag{A.7}$$

If U_1 and U_2 are unitary, then U_1U_2 is unitary:

$$(U_1U_2)^{\dagger} = U_2^{\dagger}\, U_1^{\dagger} = U_2^{-1}\, U_1^{-1} = (U_1U_2)^{-1} \quad . \tag{A.8}$$

If H is hermitian and U is unitary, then UHU^{-1} is hermitian:

$$(UHU^{-1})^{\dagger} = (U^{-1})^{\dagger} H^{\dagger} U^{\dagger} = UHU^{-1} \quad . \tag{A.9}$$

If H is hermitian then e^{iH} is unitary:

$$(e^{iH})^{\dagger} = e^{-iH\dagger} = e^{-iH} \quad . \tag{A.10}$$

If H is hermitian, then $\frac{1 + iH}{1 - iH}$ is unitary:

$$\left(\frac{1 + iH}{1 - iH}\right)^{\dagger} = \frac{1 - iH}{1 + iH} = \left(\frac{1 + iH}{1 - iH}\right)^{-1} \quad . \tag{A.11}$$

If the commutator (A,B) of two hermitian matrices is C, then $C^{\dagger} = -C$, i.e., C is antihermitian:

$$AB - BA = C$$

$$(AB-BA)^{\dagger} = B^{\dagger}A^{\dagger} - A^{\dagger}B^{\dagger} = BA-AB \quad .$$

Therefore

$$C^{\dagger} = -C \tag{A.12}$$

and if C is a number, it must be pure imaginary.

The eigenvalues of a hermitian matrix are real and those of a unitary matrix are of unit absolute value. The eigenvectors of both kinds of matrix are orthogonal.

Proof:

$$A|a'\rangle = a'|a'\rangle$$

and

$$\langle a'|A^{\dagger} = \langle a'|(a')^* \quad .$$

Therefore

$$\langle a'|A^{\dagger}|a''\rangle = \langle a'|a''\rangle(a')^*$$

$$\langle a'|A|a''\rangle = \langle a'|a''\rangle\, a'' \quad .$$

If A is hermitian, $A = A^{\dagger}$, and

$$[a'' - (a')^*]\, \langle a'|a''\rangle = 0 \quad .$$

If A is unitary, $A^{-1} = A^{\dagger}$, and

$$[(a'')^{-1} - (a')^*]\, \langle a'|a''\rangle = 0 \quad .$$

If $a' = a''$, $\langle a'|a''\rangle \neq 0$ and then

$$a' = (a')^* \qquad \text{in the hermitian case,}$$

$$(a')^{-1} = (a')^* \qquad \text{in the unitary case.}$$

In both cases if $a' \neq a''$, $\langle a'|a''\rangle = 0$.

APPENDIX B

REPRESENTATION OF LINEAR OPERATORS AS MATRICES

a) Vector and Function Spaces

Consider the set of all functions that may be expressed as a linear combination of some basic set:

$$f(x) = \sum f_k \phi_k(x) \quad . \tag{B.1}$$

The set $\phi_k(x)$ is called the basis and is supposed to be a set of linearly independent functions. They may be finite or they may be infinite like a Taylor series or a Legendre series. These functions are said to span a function space, and the number of linearly independent functions is the dimensionality of the function space.

Every function $f(x)$ may then be represented by a vector

$\{f_k\}$ and

$$a\, f(x) + b\, g(x) \leftrightarrow \{a\, f_k + b\, g_k\} \quad . \tag{B.2}$$

The set of functions f(x) is closed under addition and multiplication by a number and therefore is a vector space. The argument of f(x) is supposed to stand for $x_1 \cdots x_n$.

The scalar product may be introduced as follows:

$$(f,g) = \int \cdots \int f^*(x_1 \cdots x_n)\, g(x_1 \cdots x_n)\, \rho(x)\, dx_1 \cdots dx_n \tag{B.3}$$

where $\rho(x)$ is a weight function and where the domain of integration must be specified to complete the definition. Here f^* is the complex conjugate of f.

Then

$$(f,g) = (g,f)^* \quad . \tag{B.4}$$

We may choose the ϕ_k to be orthonormal without loss of generality:

$$(\phi_k,\, \phi_\ell) = \delta_{k\ell} \quad . \tag{B.5}$$

Then

$$(\phi_\ell,\, f) = \sum f_k(\phi_\ell,\, \phi_k) = f_\ell \quad . \tag{B.6}$$

Then also

$$(f,g) = \left(\sum f_k\phi_k, \sum g_\ell\phi_\ell\right)$$

$$= \sum f_k^* g_\ell(\phi_k, \phi_\ell) = \sum f_k^* g_k \tag{B.7}$$

which is the usual formula for the scalar product in component form.

b) Linear Operator

A is a linear operator if

$$A(f_1 + f_2) = Af_1 + Af_2 \tag{B.8}$$

$$A(cf) = c\, Af \quad . \tag{B.9}$$

(If $Af = f^2$, neither statement is true.)

Then by (B.8) and (B.9),

$$Af = A \sum f_k\phi_k = \sum Af_k\phi_k = \sum f_k(A\phi_k) \quad . \tag{B.10}$$

If

$$Af = g , \tag{B.11}$$

then

$$\sum g_\ell\phi_\ell = \sum f_k A\phi_k$$

$$\sum g_\ell(\phi_m, \phi_\ell) = \sum f_k(\phi_m, A\phi_k)$$

or

$$g_m = \sum A_{mk} f_k \tag{B.12}$$

where

$$A_{mk} = (\phi_m, A\phi_k) \quad . \tag{B.13}$$

By equations (B.12) and (B.13) one is able to replace the operator equation (B.11) by a matrix equation. The matrix (B.13) is finite (infinite) if the function space is finite (infinite).

Examples of linear operators:

$$g(x) = \left\{a \frac{d^2}{dx^2} + b \frac{d}{dx} + c\right\} f(x)$$

$$g(x) = \int K(x,x') f(x') dx'$$

for suitably restricted $\{f(x)\}$.

c) Adjoint Operators

A and B are adjoint with respect to a set of functions if

$$(\psi, A\chi) = (B\psi, \chi) \tag{B.14}$$

for any two functions in the set. We then write

$$A = B^{\dagger} \quad . \tag{B.14a}$$

A is called hermitian if $A = A^\dagger$. Then

$$(\phi_k, A\phi_\ell) = (A\phi_k, \phi_\ell) = (\phi_\ell, A\phi_k)^*$$

or

$$A_{k\ell} = A^*_{\ell k} \quad . \tag{B.15}$$

A is called unitary if

$$A^{-1} = A^\dagger \quad . \tag{B.16}$$

Then

$$(\phi_k, A^{-1}\, \phi_\ell) = (A\phi_k, \phi_\ell) = (\phi_\ell, A\phi_k)^*$$

and

$$(A^{-1})_{k\ell} = (A_{\ell k})^* \quad . \tag{B.17}$$

Therefore, hermitian (unitary) operators correspond to hermitian (unitary) matrices.

Note that if U is unitary,

$$(U\phi_k, U\phi_\ell) = (U^\dagger U\phi_k, \phi_\ell) = (\phi_k, \phi_\ell) \quad . \tag{B.18}$$

d) Choice of Basis Functions

The orthonormal basis referred to in this appendix may be chosen as the set of all normalized eigenfunctions of a hermitian or unitary operator. Alternatively one may orthonormalize a set of linearly independent functions. For example, if one orthonormalizes the set 1, x, x^2, $\cdots$, over the interval $(-1, +1)$ with the weight function $\rho(x) = 1$, then one obtains the Legendre functions. If one chooses instead the weight function $\rho(x) = e^{-x^2}$ and the interval $(-\infty, +\infty)$, then one obtains the Hermite polynomials. By choosing other intervals and weight functions, one obtains other orthonormal sets.

e) Completeness

One may approximate an arbitrary continuous function $f(x)$ in an N-dimensional function space by the method of least squares. That is, one minimizes

$$\Delta_N = (f - F_N, f - F_N) = \int |f - F_N|^2 \rho(x)\, dx \qquad \text{(B.19)}$$

when $F_N(x)$ is the approximation function. It is easily shown that Δ_N is a minimum when

$$F_N(x) = \sum_1^N f_k \phi_k \qquad \text{(B.20)}$$

where

$$f_k = (\phi_k, f) \quad . \qquad \text{(B.20a)}$$

Then also

$$(f,f) \geq \sum_{1}^{N} |f_k|^2 \quad . \tag{B.21}$$

If one now passes to the case of an infinite dimensional functional space, and if

$$\lim_{N\to\infty} \Delta_N = 0 \quad , \tag{B.22}$$

then one says that $F_N(x)$ converges to $f(x)$ in the mean and that

$$(f,f) = \sum_{1}^{\infty} |f_k|^2 \quad . \tag{B.23}$$

This basis is then said to be complete.

f) Dirac Notation

The relation to the Dirac bracket notation is the following:

$$\langle a'|A|a''\rangle = (\phi_{a'}, A\phi_{a''})$$
$$= \int \cdots \int \phi^*_{a'}(x_1 \cdots x_n)\, A\phi_{a''}(x_1 \cdots x_n)\, \rho(x_1 \cdots x_n)\, dx_1 \cdots dx_n .$$

Equation (B.23), which expresses completeness, reads in Dirac notation as follows:

$$< | > = \sum_{a'} < |a'> <a'| > \quad , \tag{B.24}$$

where $| >$ and $<a'| >$ correspond to f and f_k. Equation (B.24) expresses the expectation value, for an arbitrary state, of the operator relation

$$\sum_{a'} |a'> <a'| = 1 \tag{B.25}$$

which is also (3.14) of Chapter 2.

Notice also that (B.25) implies

$$\sum_{a'} <x'|a'> <a'|x''> = <x'|x''> \tag{B.26}$$

or

$$\sum_{a'} \phi_{a'}(x') \phi_{a'}(x'')^* = \delta(x'-x'') \quad . \tag{B.26a}$$

Similarly

$$\begin{aligned} <x'|A|x''> &= \sum <x'|a'> <a'|A|a''> <a''|x''> \\ &= \sum a' <x'|a'> <a'|x''> \\ &= \sum a' \phi_{a'}(x') \phi_{a'}(x'')^* \end{aligned} \tag{B.27}$$

so that the operator relation

$$A |f> = |g>$$

may be expressed as an integral equation as follows

$$\int <x'|A|x''> dx' <x''|f> = <x'|g>$$

or

$$\int \langle x'|A|x''\rangle \, dx'' \, f(x'') = g(x') \tag{B.28}$$

where the kernel of the equation is given by (B.27). Then (B.28) may be regarded as the generalization of (B.12) to a continuous matrix.

Reference:

Courant and Hilbert, Methods of Mathematical Physics, Volume I, Interscience 1953.

APPENDIX C

SPIN ALGEBRA AND VECTOR ANALYSIS

The three Pauli matrices are

$$\sigma_1 = \begin{pmatrix} 0 & 1 \\ 1 & 0 \end{pmatrix} \qquad \sigma_2 = \begin{pmatrix} 0 & -i \\ i & 0 \end{pmatrix} \qquad \sigma_3 = \begin{pmatrix} 1 & 0 \\ 0 & -1 \end{pmatrix}$$

and they satisfy the following relations:

$$\sigma_k \sigma_\ell - \sigma_\ell \sigma_k = 2i\varepsilon_{k\ell m}\sigma_m$$

$$\sigma_k \sigma_\ell + \sigma_\ell \sigma_k = 2\delta_{k\ell}$$

$$\sigma_k \sigma_\ell = i\varepsilon_{k\ell m}\sigma_m + \delta_{k\ell} \qquad .$$

We may associate a vector a_k with a traceless matrix A as follows:

$$a_k = \mathrm{Tr}\, A\sigma_k \ , \qquad A = \sum a_k \sigma_k \qquad .$$

Here Tr is an abbreviation for $\frac{1}{2}$ x trace.

Let $\underset{\sim}{a}$ and $\underset{\sim}{b}$ be two vectors and let A and B be the associated matrices as follows:

$$A = \underset{\sim}{a}\underset{\sim}{\sigma} \quad , \qquad B = \underset{\sim}{b}\underset{\sim}{\sigma} \quad .$$

Then the anticommutator of A and B is

$$\{A, B\} = AB + BA = \sum a_m b_n (\sigma_m \sigma_n + \sigma_n \sigma_m) = 2\underset{\sim}{a}\underset{\sim}{b} \quad .$$

The commutator is

$$(A, B) = AB - BA = \sum a_m b_n (\sigma_m, \sigma_n) = 2i \sum \varepsilon_{mnp} \, a_m b_n \sigma_p$$

or

$$(A, B) = 2i(\underset{\sim}{a} \times \underset{\sim}{b}) \cdot \underset{\sim}{\sigma} \quad .$$

Therefore

$$AB = \underset{\sim}{a}\underset{\sim}{b} + i(\underset{\sim}{a} \times \underset{\sim}{b}) \cdot \underset{\sim}{\sigma}$$

or the matrix product carries both the inner and outer products of the associated vectors.

One may show

$$\mathrm{Tr}\, AB = \underset{\sim}{a}\underset{\sim}{b}$$

$$\mathrm{Tr}\, ABC = i(\underset{\sim}{a} \times \underset{\sim}{b}) \cdot \underset{\sim}{c}$$

$$\big((A, B), C\big) = (2i)^2 \big((\underset{\sim}{a} \times \underset{\sim}{b}) \times \underset{\sim}{c}\big) \cdot \underset{\sim}{\sigma} \quad .$$

APPENDIX D

ALGEBRA OF REPRESENTATION THEORY

In Chapter 2, section (2.8), it is claimed that a complete set of commuting observables may be derived from the symmetry group by a general construction that is there deferred to the appendix. This construction will now be described and also some additional background information will be presented in order to supplement the discussion in Chapters 4 and 5 that depends in part on the representation theory of groups.

a) Complete Sets of Commuting Observables

To describe the construction of a complete set of commuting observables for a given symmetry group we shall first state Schur's lemma.[1]

Let

$$[M, D(P)] = 0 \qquad P \varepsilon G \tag{D.1}$$

for every number P of the group G. If D(P) is an irreducible representation of G then Schur's lemma states that M is a multiple of the unit matrix:

$$M = mD(I) \quad . \tag{D.2}$$

As an example of M consider

$$\Lambda_A = \frac{1}{g} \sum_{P\varepsilon G} D(P)\ D(A)\ D(P)^{-1} \qquad A\varepsilon G \tag{D.3}$$

where it is now assumed that G is a finite group with g elements; A is any element of G; and the sum is over the complete group. Then, if X is also an arbitrary element of G, we have

$$\begin{aligned} D(X)\ \Lambda_A\ D(X)^{-1} &= \frac{1}{g} \sum_{P\varepsilon G} D(XP)\ D(A)\ D(XP)^{-1} \\ &= \frac{1}{g} \sum_{Y\varepsilon G} D(Y)\ D(A)\ D(Y)^{-1} \end{aligned}$$

$$D(X)\ \Lambda_A\ D(X)^{-1} = \Lambda_A \tag{D.4}$$

or

$$D(X)\ \Lambda_A = \Lambda_A\ D(X) \quad . \tag{D.5}$$

Then by Schur's lemma

$$\Lambda_A = \lambda_A\ D(I) \quad . \tag{D.6}$$

Notice also that

$$\Lambda_A = \frac{1}{g} \sum_{P \varepsilon G} D(PAP^{-1}) \quad . \qquad (D.7)$$

The elements PAP^{-1} obtained by letting P run over the complete group are not all different, but all the distinct elements so obtained are repeated the same number of times. These distinct elements are said to form a class. Denote the class corresponding to A by C_α, and let the number of elements in the class be n_α. Then (D.7) may also be written

$$\Lambda_\alpha \equiv \Lambda_A = \frac{1}{n_\alpha} \sum_{Z \varepsilon C_\alpha} D(Z) \quad . \qquad (D.8)$$

By construction all elements of this class are conjugate to A and therefore to each other. Of course A is not distinguished in any way, and any other member of this class could be used to generate C_α.

The complete group is composed of non-overlapping classes

$$G = \bigcup_\alpha C_\alpha \quad . \qquad (D.9)$$

(The number of classes is important for representation theory and in fact the number of irreducible representations equals the number of classes.)

By (D.5) the Λ_α commute with all group elements and therefore they also commute among themselves

$$(\Lambda_\alpha, \Lambda_\beta) = 0 \quad . \tag{D.10}$$

The Λ_α form a maximal commuting set and their eigenvalues (λ_α) may be chosen as quantum numbers as explained in Chapter 2, section (2.8), if G is a symmetry group.

b) Orthogonality Relations

The construction (D.3) may be generalized as follows. Define

$$M = \sum_{P \varepsilon G} D^\ell(P)\, Q\, D^{\ell'}(P)^{-1} \tag{D.11}$$

where Q is an arbitrary matrix that does not necessarily correspond to an element of the group, and where $\{D^\ell(R)\}$ and $\{D^{\ell'}(R)\}$ are any two irreducible representations of the group. The original construction (D.3) is recovered by putting Q = D(A) and $\ell = \ell'$. Just as before, however,

$$D^\ell(X)\, M\, D^{\ell'}(X)^{-1} = \sum_{P \varepsilon G} D^\ell(XP)\, Q\, D^{\ell'}(XP)^{-1}$$

and

$$D^\ell(X)\, M\, D^{\ell'}(X)^{-1} = M \quad .$$

In this case the conclusion of Schur's lemma may be generalized to the following theorem, that is also stated without proof.[2]

If ℓ and ℓ' are two irreducible representations, then either M is the null matrix, or a matrix with a non-vanishing determinant. In the latter case M has an inverse and the two irreducible representations, ℓ and ℓ', are equivalent. If they are equivalent, take them to be the same. Then

$$\sum_{P\varepsilon G} D^{\ell}(P)\; Q\; D^{\ell'}(P^{-1}) = m(Q)\; \delta^{\ell\ell'}\; D^{\ell}(I) \tag{D.13}$$

by combining Schur's lemma with the preceding statement.

Since Q is arbitrary, the above matrix equation has ℓ^2 components that can be made more explicit by introducing the following basis matrices Q(a,b):

$$\left(Q(a,b)\right)_{\lambda\mu} = \delta_{\lambda}{}^{a}\; \delta_{\mu}{}^{b} \quad . \tag{D.14}$$

By substituting (D.14) in (D.13) one finds

$$\sum_{P}\sum_{\lambda\mu} D^{\ell}_{\alpha\lambda}(P)\; (\delta_{\lambda}{}^{a}\; \delta_{\mu}{}^{b})\; D^{\ell'}_{\mu\beta}(P^{-1}) = m(a,b)\; \delta^{\ell\ell'}\; D^{\ell}_{\alpha\beta}(I) \quad .$$

Put $\alpha = \beta$ and sum. Then

$$\sum_{\beta}\sum_{P} D^{\ell}_{\beta a}(P)\; D^{\ell'}_{b\beta}(P^{-1}) = m(a,b)\; \delta^{\ell\ell'} \sum_{\beta} D^{\ell}_{\beta\beta}(I)$$

or

$$\sum_{P} D_{ba}(I) = m(a,b)\; d_{\ell}$$

where d_ℓ is the dimensionality of the matrices $D^\ell(P)$. The preceding relation then implies

$$m(a,b) = \frac{g}{d_\ell}\,\delta(a,b) \tag{D.15}$$

where g is the number of elements of G.

Finally substituting (D.15) back in (D.13) one finds

$$\sum_P D^\ell_{\alpha a}(P)\, D^{\ell'}_{\beta b}(P)^* = \frac{g}{d_\ell}\,\delta_{ab}\,\delta_{\alpha\beta}\,\delta^{\ell\ell'} \tag{D.16}$$

since we are always assuming that the matrices D(P) are unitary.

These are the group orthogonality relations. To extend these formulas to continuous groups, it is necessary to introduce a measure on the group and replace the sum by an integral, as in Chapter 5 and in the following appendix.

c) Reduction of a Reducible Representation

Putting $\alpha = a$ and $\beta = b$, and summing, one also finds

$$\sum_P \chi^\ell(P)^*\,\chi^{\ell'}(P) = g\,\delta^{\ell\ell'} \tag{D.17}$$

where

$$\chi^\ell(P) = \sum_a D^\ell_{aa}(P) \tag{D.18}$$

is the trace of $D^{\ell}(P)$. All matrices belonging to elements of the same class have the same trace, since such matrices are related by similarity transformations. This set of traces, one for each class, is called the character of the representation, irreducible or not. The characters of the irreducible representations are orthogonal in the sense of equation (D.17). Notice that (D.17) can also be written

$$\sum_{\alpha} n_{\alpha} (\chi_{\alpha}{}^{\ell})^{*} (\chi_{\alpha}{}^{\ell'}) = g\, \delta^{\ell\ell'} \qquad \text{(D.17a)}$$

where $\chi_{\alpha}{}^{\ell}$ is the trace associated with the class C_{α} and n_{α} is the number of elements in this class.

If $\{D(P)\}$ is a reducible representation, it may be reduced as follows:

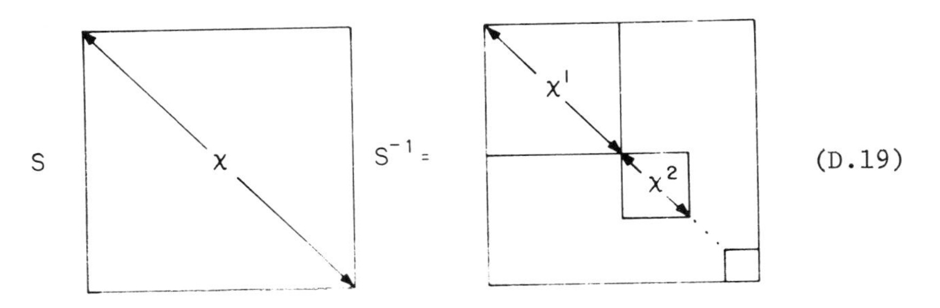

(D.19)

Clearly

$$\chi(P) = \sum c_{\ell}\, \chi^{\ell}(P) \qquad \text{(D.20)}$$

where c_{ℓ} is the number of times the irreducible representation

D^{ℓ} appears on the right. But this equation may be solved for the c_{ℓ} as follows:

$$c_{\ell} = \frac{1}{g} \sum_{P} \chi^{\ell}(P)^* \chi(P) \tag{D.21}$$

by making use of (D.17).

Therefore one is able to reduce an arbitrary representation by solving for the c_{ℓ} provided that one knows the characters of the irreducible representations.

The new reduced basis may be obtained by use of the following projection operators

$$e_{\ell} = \frac{d_{\ell}}{g} \sum_{P} \chi_{\ell}(P)^* D(P) \tag{D.22}$$

$$= \frac{d_{\ell}}{g} \sum_{\alpha} (\chi_{\ell}{}^{\alpha})^* \sum_{P\varepsilon C_{\alpha}} D(P)$$

$$= \frac{d_{\ell}}{g} \sum_{\alpha} (\chi_{\ell}{}^{\alpha})^* n_{\alpha}\Lambda_{\alpha} \tag{D.22a}$$

where $\chi_{\ell}{}^{\alpha}$ belongs to the α-class of the ℓth irreducible representation. Therefore the e_{ℓ} are sums of the class functions Λ_{α} and therefore the e_{ℓ} also commute among themselves. One may easily show that they are also projection operators:

$$e_{\ell}\, e_{\ell'} = e_{\ell}\, \delta_{\ell\ell'} \tag{D.23a}$$

$$\sum e_\ell = I \quad . \tag{D.23b}$$

These operators project out states ψ_ℓ that correspond to irreducible representations and between which there are no elements of the S-matrix. For

$$(\psi_\ell,\ S\psi_{\ell'}) = (e_\ell\psi,\ S\ e_{\ell'}\psi) = (\psi,\ e_\ell\ S\ e_{\ell'}\psi)$$

$$= (\psi,\ S\ e_\ell e_{\ell'}\psi) = (\psi,\ S\ e_\ell\psi)\delta_{\ell\ell'} \tag{D.24}$$

since G is a symmetry group and therefore $(S,\ e_\ell) = 0$. The ψ_ℓ may be labeled by the quantum numbers that are defined by the complete commuting set $\{\Lambda_\alpha\}$. According to (D.24) there are no transitions between these states.

References

1. See Wigner, page 75, Theorem 2.
2. Ibid., page 76, Theorem 3.

APPENDIX E

DIFFERENTIAL OPERATORS IN THE GROUP SPACE

The mathematics of Appendix E underlies much of this book, since it provides a connection between the representation of the generators of a group as Heisenberg matrices, on the one hand, and as Schrödinger differential operators, on the other hand. It is not necessary, however, to work through this Appendix in order to read the main text. We shall here limit our considerations to the rotation group.

a) Left and Right Generators and Integrability Condition

Let us begin with (3.10) of Chapter 4, namely,

$$X = U \overset{\circ}{X} U^{-1} \tag{E.1}$$

where $\det U = 1$ and $U^{+} = U^{-1}$. Here $\overset{\circ}{X}$ is some given hermitian, traceless, 2×2 matrix, and U is a two dimensional representation of the rotation group. Then U and X depend on three parameters that characterize the rotation. These may be (α,β,γ) or $(\omega_1,\omega_2,\omega_3)$ or some other set, and will be denoted by (a^1,a^2,a^3). Then

$$\partial_\alpha X = (\partial_\alpha U)\, U^{-1} X + X\, U\, \partial_\alpha U^{-1} = (X,\, U\, \partial_\alpha U^{-1})$$

where

$$\partial_\alpha = \partial/\partial a^\alpha$$

and

$$a^\alpha = (a^1,a^2,a^3) \quad .$$

Then

$$\partial_\alpha X = \frac{1}{i}\left(X,\, \Lambda_\alpha(+)\right) \tag{E.2+}$$

where $\Lambda_\alpha(+)$ is the hermitian matrix:

$$\Lambda_\alpha(+) = i\, U\, \partial_\alpha U^{-1} \quad . \tag{E.3+}$$

Then also

$$\partial_\alpha U = i\, \Lambda_\alpha(+)\, U \quad . \tag{E.4+}$$

We next introduce $\Lambda_\alpha(-)$ by the definition:

$$\Lambda_\alpha(-) = -\, i\, U^{-1}\, \partial_\alpha U \quad . \tag{E.3-}$$

Then

$$\partial_\alpha U = i\, U\, \Lambda_\alpha(-) \quad . \tag{E.4-}$$

The $\Lambda_\alpha(\pm)$ may be called left and right generators. The notation $(\pm)$ refers for example to (E.4±). Notice also that

$$U\, \Lambda_\alpha(-)\, U^{-1} = \Lambda_\alpha(+) \quad . \tag{E.5}$$

Since equations (E.4±) can be integrated to yield U, they must satisfy conditions of integrability, namely:

$$\partial_\beta \partial_\alpha U = \partial_\alpha \partial_\beta U$$

or

$$\partial_\beta \left(i\, \Lambda_\alpha(+)\, U\right) = \partial_\alpha \left(i\, \Lambda_\beta(+)\, U\right)$$

$$i[\partial_\beta \Lambda_\alpha(+) - \partial_\alpha \Lambda_\beta(+)]\, U = [-\Lambda_\beta(+)\Lambda_\alpha(+) + \Lambda_\alpha(+)\Lambda_\beta(+)]\, U$$

and therefore

$$\partial_\beta \Lambda_\alpha(+) - \partial_\alpha \Lambda_\beta(+) + i\left(\Lambda_\alpha(+),\, \Lambda_\beta(+)\right) = 0 \quad . \tag{E.6+}$$

By the same argument one finds

$$\partial_\beta \Lambda_\alpha(-) - \partial_\alpha \Lambda_\beta(-) - i\left(\Lambda_\alpha(-),\, \Lambda_\beta(-)\right) = 0 \quad . \tag{E.6-}$$

b) The Group Space

The matrix U depends on the three parameters or coordinates a^α.

If

$$U = e^{\frac{i}{2} \underset{\sim}{\sigma} \underset{\sim}{\omega}}$$

then $(a^1, a^2, a^3) = (\omega^1, \omega^2, \omega^3)$ and every point contained in a sphere of radius π represents a possible rotation. This complete domain is the group space. Then U and all derived matrices like $\Lambda_\mu(\pm)$ are defined over this space.

Since det U = 1, it follows that

$$\mathrm{Tr}\ \Lambda_\alpha(\pm) = 0 \quad . \tag{E.7}$$

Therefore $\Lambda_\alpha(\pm)$ may be written as a linear combination of traceless matrices:

$$\Lambda_\mu(\pm) = R_o^{-1} \sum \lambda_\mu{}^i(\pm)\ \sigma_i \tag{E.8}$$

where the σ_i are the Pauli matrices, R_o is an undetermined constant, and the $\lambda_\mu{}^i(\pm)$ are the following functions:

$$\lambda_\mu{}^i(\pm) = R_o\ \mathrm{Tr}\ \Lambda_\mu(\pm)\ \sigma^i \quad . \tag{E.9}$$

The $\lambda_\mu{}^i(+)$ and $\lambda_\mu{}^i(-)$ are then two triads of vector fields defined over the group space.

The integrability conditions (E.6±) may be expressed in terms of the fields $\lambda_\mu{}^i(\pm)$ as follows:

$$\partial_\alpha \lambda_\beta{}^s(\pm) - \partial_\beta \lambda_\alpha{}^s(\pm) \pm \frac{2}{R_o} \varepsilon^s_{mn} \lambda_\alpha{}^m(\pm)\, \lambda_\beta{}^n(\pm) = 0. \quad \text{(E.10±)}$$

c) Geometry

We may define reciprocal vector fields $\lambda_i{}^\mu(\pm)$ by the relations

$$\lambda_\mu{}^i(\pm)\, \lambda_k{}^\mu(\pm) = \delta_k{}^i \quad \text{(E.11)}$$

$$\lambda_\mu{}^i(\pm)\, \lambda_i{}^\sigma(\pm) = \delta_\mu{}^\sigma \quad . \quad \text{(E.12)}$$

The reciprocal fields thus define the matrix $||\lambda_\mu{}^i||$ inverse to $||\lambda_i{}^\mu||$.

Next define the connections

$$L^\mu{}_{\alpha\beta}(\pm) = \lambda_i{}^\mu(\pm)\, \partial_\beta\, \lambda_\alpha{}^i(\pm) \quad \text{(E.13)}$$

$$= - \partial_\beta\, \lambda_i{}^\mu(\pm)\, \lambda_\alpha{}^i(\pm) \quad \text{(E.14)}$$

by (E.12). Then

$$\partial_\beta\, \lambda_k{}^\mu(\pm) + L^\mu{}_{\alpha\beta}(\pm)\, \lambda_k{}^\alpha(\pm) = 0 \quad \text{(E.15)}$$

$$\partial_\beta\, \lambda_\mu{}^k(\pm) - L^\sigma{}_{\mu\beta}(\pm)\, \lambda_\sigma{}^k(\pm) = 0 \quad . \quad \text{(E.16)}$$

The geometric interpretation of (E.15) is that the fields $\lambda_k^{\ \mu}(\pm)$ are "parallel" with respect to connections $L^{\mu}_{\ \alpha\beta}(\pm)$. Equations (E.15) and (E.16) differ only in that the latter describes covariant components.

Next define the metric

$$g_{\alpha\beta}(\pm) = \sum_i \lambda_\alpha^{\ i}(\pm)\ \lambda_\beta^{\ i}(\pm) \qquad . \tag{E.17$\pm$}$$

Then

$$g_{\alpha\beta}(\pm) = R_o^{\ 2}\ \mathrm{Tr}\ \Lambda_\alpha(\pm)\ \Lambda_\beta(\pm) = g_{\beta\alpha}(\pm) \qquad . \tag{E.18$\pm$}$$

But

$$\begin{aligned} g_{\alpha\beta}(-) &= R_o^{\ 2}\ \mathrm{Tr}\ \Lambda_\alpha(-)\ \Lambda_\beta(-) \\ &= R_o^{\ 2}\ \mathrm{Tr}\ \Lambda_\alpha(+)\ \Lambda_\beta(+) = g_{\alpha\beta}(+) \end{aligned}$$

by (E.5). Therefore

$$g_{\alpha\beta}(-) = g_{\alpha\beta}(+) = g_{\alpha\beta} = g_{\beta\alpha} \qquad . \tag{E.19}$$

From $g_{\alpha\beta}$ one may obtain $g^{\alpha\beta}$ in the usual way. Notice that

$$g_{\alpha\beta}\ \lambda_k^{\ \beta} = \left[\sum_i \lambda_\alpha^{\ i}\ \lambda_\beta^{\ i}\right]\ \lambda_k^{\ \beta} = \lambda_\alpha^{\ k} \tag{E.20a}$$

$$g^{\gamma\alpha}\ \lambda_\alpha^{\ k} = \lambda_k^{\ \gamma} \qquad . \tag{E.20b}$$

As a matrix equation (E.13) reads

$$L^{\mu}{}_{\alpha\beta}(\pm) = R_o{}^2 \operatorname{Tr} \Lambda^{\mu}(\pm)\, \partial_{\beta}\, \Lambda_{\alpha}(\pm) \quad . \tag{E.21$\pm$}$$

Let $T^{\mu}{}_{\alpha\beta}$ be the antisymmetric part of $L^{\mu}{}_{\alpha\beta}$:

$$T^{\mu}{}_{\alpha\beta} = \frac{1}{2}\,(L^{\mu}{}_{\alpha\beta} - L^{\mu}{}_{\beta\alpha}) \quad . \tag{E.22}$$

Then

$$\begin{aligned} T^{\mu}{}_{\alpha\beta}(\pm) &= R_o{}^2\, \frac{1}{2} \operatorname{Tr} \Lambda^{\mu}(\pm)\, [\partial_{\beta}\Lambda_{\alpha}(\pm) - \partial_{\alpha}\Lambda_{\beta}(\pm)] \\ &= \pm\, \frac{i}{2}\, R_o{}^2 \operatorname{Tr} \Lambda^{\mu}(\pm)\, \big(\Lambda_{\beta}(\pm),\, \Lambda_{\alpha}(\pm)\big) \end{aligned} \tag{E.23$\pm$}$$

by (E.6±). Therefore

$$\begin{aligned} T^{\mu}{}_{\alpha\beta}(+) &= \frac{i}{2}\, R_o{}^2 \operatorname{Tr} \Lambda^{\mu}(+)\, \big(\Lambda_{\beta}(+),\, \Lambda_{\alpha}(+)\big) \\ &= \frac{i}{2}\, R_o{}^2 \operatorname{Tr} \Lambda^{\mu}(-)\, \big(\Lambda_{\beta}(-),\, \Lambda_{\alpha}(-)\big) \end{aligned} \tag{E.24}$$

by (E.5). Therefore

$$T^{\mu}{}_{\alpha\beta}(+) = -\, T^{\mu}{}_{\alpha\beta}(-) \tag{E.25}$$

by (E.23±), or

$$T^{\mu}{}_{\beta\alpha}(+) = T^{\mu}{}_{\alpha\beta}(-) \quad . \tag{E.26}$$

Finally by (E.8) and (E.24)

$$T^{\mu}{}_{\alpha\beta}(\pm) = \pm \frac{1}{R_o} \varepsilon^{p}{}_{mn} \lambda_{p}{}^{\mu}(\pm) \lambda_{\alpha}{}^{m}(\pm) \lambda_{\beta}{}^{n}(\pm) \qquad \text{(E.27a,}\pm\text{)}$$

and

$$T_{\gamma\alpha\beta}(\pm) = g_{\gamma\mu} T^{\mu}{}_{\alpha\beta}(\pm)$$

$$= \pm \frac{1}{R_o} \varepsilon_{pmn} \lambda_{\gamma}{}^{p}(\pm) \lambda_{\alpha}{}^{m}(\pm) \lambda_{\beta}{}^{n}(\pm) \ . \qquad \text{(E.27b,}\pm\text{)}$$

It follows that the tensor $T_{\gamma\alpha\beta}(\pm)$ is completely antisymmetric.

Let the symmetric part of the connection be

$$\Gamma^{\mu}{}_{\alpha\beta} = \frac{1}{2}(L^{\mu}{}_{\alpha\beta} + L^{\mu}{}_{\beta\alpha}) \qquad \text{(E.28)}$$

so that

$$L^{\mu}{}_{\alpha\beta} = \Gamma^{\mu}{}_{\alpha\beta} + T^{\mu}{}_{\alpha\beta} \ .$$

By calculating

$$\partial_{\gamma} g_{\alpha\beta} = \partial_{\gamma}\left[\sum_i \lambda_{\alpha}{}^{i}(\pm) \lambda_{\beta}{}^{i}(\pm)\right]$$

with the aid of (E.16), we may show that

$$\partial_{\gamma} g_{\alpha\beta} - L^{\sigma}{}_{\alpha\gamma}(\pm) g_{\sigma\beta} - L^{\sigma}{}_{\beta\gamma}(\pm) g_{\alpha\sigma} = 0$$

and therefore, since $T_{\mu\alpha\beta}$ is completely antisymmetric by

(E.27b),

$$\partial_\gamma g_{\alpha\beta} - \Gamma^\sigma{}_{\alpha\gamma}(\pm)\, g_{\sigma\beta} - \Gamma^\sigma{}_{\beta\gamma}(\pm)\, g_{\alpha\sigma} = 0 \qquad .$$

This equation may be solved in the usual way:

$$\Gamma^\sigma{}_{\alpha\beta}(\pm) = \frac{1}{2}\, g^{\sigma\lambda} \left(\frac{\partial g_{\lambda\beta}}{\partial a^\alpha} + \frac{\partial g_{\lambda\alpha}}{\partial a^\beta} - \frac{\partial g_{\alpha\beta}}{\partial a^\lambda} \right) \qquad . \qquad \text{(E.29)}$$

Therefore

$$\Gamma^\sigma{}_{\alpha\beta}(+) = \Gamma^\sigma{}_{\alpha\beta}(-) \qquad ; \qquad \text{(E.30)}$$

(E.30) and (E.26) imply

$$L^\mu{}_{\alpha\beta}(+) = L^\mu{}_{\beta\alpha}(-) \qquad . \qquad \text{(E.31)}$$

d) Algebra of Displacement Operators

The underlying vector fields $\left(\lambda_m{}^\mu(\pm)\right)$ define the displacement operators

$$X_m(\pm) = \frac{1}{i}\, \lambda_m{}^\mu(\pm)\, \partial_\mu \qquad . \qquad \text{(E.32)}$$

These obey the commutation rules

$$\left(X_m(\pm),\ X_n(\pm)\right) = -\left(\lambda_m{}^{\mu}(\pm)\ \partial_\mu,\ \lambda_n{}^{\sigma}(\pm)\ \partial_\sigma\right)$$

$$= [-\lambda_m{}^{\mu}(\pm)\ \left(\partial_\mu \lambda_n{}^{\sigma}(\pm)\right)\ \partial_\sigma + \left(\partial_\sigma\ \lambda_m{}^{\mu}(\pm)\right)\ \lambda_n{}^{\sigma}(\pm)\ \partial_\mu]$$

$$= [-\lambda_m{}^{\sigma}(\pm)\ \partial_\sigma\ \lambda_n{}^{\mu}(\pm) + \lambda_n{}^{\sigma}(\pm)\ \partial_\sigma\ \lambda_m{}^{\mu}(\pm)]\ \partial_\mu$$

$$= [L^{\mu}{}_{\alpha\sigma}(\pm)\ \lambda_n{}^{\alpha}(\pm)\ \lambda_m{}^{\sigma}(\pm) - L^{\mu}{}_{\alpha\sigma}(\pm)\ \lambda_m{}^{\alpha}(\pm)\ \lambda_n{}^{\sigma}(\pm)]\ \partial_\mu$$

$$= [L^{\mu}{}_{\alpha\sigma}(\pm) - L^{\mu}{}_{\sigma\alpha}(\pm)]\ \lambda_n{}^{\alpha}(\pm)\ \lambda_m{}^{\sigma}(\pm)\ \partial_\mu$$

$$= 2\ T^{\mu}{}_{\alpha\sigma}(\pm)\ \lambda_n{}^{\alpha}(\pm)\ \lambda_m{}^{\sigma}(\pm)\ \partial_\mu$$

$$= \pm \frac{2}{R_o}\left(\varepsilon^{p}{}_{rs}\ \lambda_p{}^{\mu}(\pm)\ \lambda_\alpha{}^{r}(\pm)\ \lambda_\beta{}^{s}(\pm)\right)\ \lambda_n{}^{\alpha}(\pm)\ \lambda_m{}^{\beta}(\pm)\ \partial_\mu$$

by (E.27±)

$$= \pm \frac{2}{R_o}\ \varepsilon^{p}{}_{nm}\ \lambda_p{}^{\mu}(\pm)\ \partial_\mu \quad .$$

Therefore

$$\left(X_m(\pm),\ X_n(\pm)\right) = \mp \frac{2}{R_o}\ i\ \varepsilon^{p}{}_{mn}\ X_p(\pm) \quad . \tag{E.33}$$

On the other hand

$$\left(X_m(+),\ X_n(-)\right) = -\left(\lambda_m{}^{\mu}(+)\ \partial_\mu,\ \lambda_n{}^{\sigma}(-)\ \partial_\sigma\right)$$

$$= -\ \lambda_m{}^{\mu}(+)\ \left(\partial_\mu \lambda_n{}^{\sigma}(-)\right)\ \partial_\sigma + \left(\partial_\sigma \lambda_m{}^{\mu}(+)\right)\ \lambda_n{}^{\sigma}(-)\ \partial_\mu$$

$$= [-\lambda_m{}^{\sigma}(+)\ \partial_\sigma\ \lambda_n{}^{\mu}(-) + \lambda_n{}^{\sigma}(-)\ \partial_\sigma\ \lambda_m{}^{\mu}(+)]\ \partial_\mu$$

$$\left(X_m(+),\ X_n(-)\right) =$$

$$= [L^{\mu}{}_{\alpha\sigma}(-)\ \lambda_n{}^{\alpha}(-)\ \lambda_m{}^{\sigma}(+) - L^{\mu}{}_{\alpha\sigma}(+)\ \lambda_m{}^{\alpha}(+)\ \lambda_n{}^{\sigma}(-)]\ \partial_\mu$$

$$= L^{\mu}{}_{\alpha\sigma}(-)\ [\lambda_n{}^{\alpha}(-)\ \lambda_m{}^{\sigma}(+) - \lambda_n{}^{\alpha}(-)\ \lambda_m{}^{\sigma}(+)]\ \partial_\mu \quad .$$

Therefore

$$\left(X_m(+),\ X_n(-)\right) = 0 \quad . \tag{E.34}$$

The algebra of the displacement operators is then described by the commutation rules:

$$\left(X_m(\pm),\ X_n(\pm)\right) = \mp \frac{2i}{R_o}\ \varepsilon^p{}_{mn}\ X_p(\pm)$$

$$\left(X_m(\pm),\ X_n(\mp)\right) = 0 \quad .$$

These are the equations of the Lie algebra belonging to the rotation group or alternatively the equations that must be satisfied by the angular momentum operators (up to a scale factor). They may be satisfied by constant matrices or by the different operators $X_m(\pm)$.

The differential operator corresponding to the total angular momentum is

$$X^2 = \sum_m X_m(\pm)\ X_m(\pm) \quad . \tag{E.35}$$

One easily finds that

$$\left(X^2,\ X_m(\pm)\right) = 0 \quad . \tag{E.36}$$

The expression for X^2 as a differential operator is

$$X^2 = \sum (\lambda_m{}^\mu\ \partial_\mu)(\lambda_m{}^\sigma\ \partial_\sigma) = g^{\mu\sigma}\ \partial_\mu\ \partial_\sigma + \sum_m \lambda_m{}^\mu(\partial_\mu\ \lambda_m{}^\sigma)\ \partial_\sigma \ .$$

The second term contains

$$\sum \lambda_m{}^\mu\ \partial_\mu\ \lambda_m{}^\sigma = - \sum \lambda_m{}^\mu\ L^\sigma{}_{\alpha\mu}\ \lambda_m{}^\alpha = - \Gamma^\sigma{}_{\alpha\mu}\ g^{\alpha\mu} \quad .$$

But

$$\partial_\alpha\ g^{\alpha\sigma} + \Gamma^\alpha{}_{\mu\alpha}\ g^{\mu\sigma} + \Gamma^\sigma{}_{\mu\alpha}\ g^{\alpha\mu} = 0 \quad .$$

Then

$$\sum \lambda_m{}^\mu\ \partial_\mu\ \lambda_m{}^\sigma = \partial_\alpha\ g^{\alpha\sigma} + \Gamma^\alpha{}_{\mu\alpha}\ g^{\mu\sigma} = \frac{1}{\sqrt{g}}\ \partial_\mu\ (\sqrt{g}\ g^{\mu\sigma})$$

since*

$$\Gamma^\alpha{}_{\mu\alpha} = \frac{1}{\sqrt{g}}\ \partial_\mu\ \sqrt{g} \quad .$$

Therefore

$$X^2 = \frac{1}{\sqrt{g}}\ \partial_\mu\ \sqrt{g}\ g^{\mu\sigma}\ \partial_\sigma \quad . \tag{E.37}$$

e) Special Coordinate Systems

The preceding constructions for $X_m(\pm)$ and X^2 hold for all parameterizations of the rotation group. We shall now compute these operators for the special cases that have been used in the text.

(1) Eulerian Angles

In this case

$$U(\alpha,\beta,\gamma) = U_3(\alpha/2)\ U_2(\beta/2)\ U_3(\gamma/2)$$

where

$$U_m(\theta) = e^{i\sigma_m\theta} .$$

We find

$$\begin{aligned}
\Lambda_\alpha(+) &= \frac{1}{2}\sigma_3 & \Lambda_\alpha(-) &= \frac{1}{2} U^{-1}\sigma_3 U \\
\Lambda_\beta(+) &= \frac{1}{2} U_3(\alpha)\sigma_2 & \Lambda_\beta(-) &= \frac{1}{2}\sigma_2 U_3(\gamma) \\
\Lambda_\gamma(+) &= \frac{1}{2} U\sigma_3 U^{-1} & \Lambda_\gamma(-) &= \frac{1}{2}\sigma_3
\end{aligned} \tag{E.38}$$

$$g_{\mu\nu} = \frac{1}{4} R_o^{\ 2} \begin{pmatrix} 1 & 0 & \cos\beta \\ 0 & 1 & 0 \\ \cos\beta & 0 & 1 \end{pmatrix} \tag{E.39}$$

$$g^{\mu\nu} = \frac{4}{R_o^{\ 2} \sin^2\beta} \begin{pmatrix} 1 & 0 & -\cos\beta \\ 0 & \sin^2\beta & 0 \\ -\cos\beta & 0 & 1 \end{pmatrix}$$

$$g^{1/2} = \frac{1}{8} R_o^{\ 3} \sin\beta \quad . \tag{E.40}$$

The various differential operators are

$$X^2 = \frac{1}{g^{1/2}} \partial_\mu \, g^{1/2} \, g^{\mu\nu} \, \partial_\nu$$

$$= g^{\mu\nu} \partial_\mu \partial_\nu + \frac{1}{g^{1/2}} \left[\frac{\partial}{\partial\beta} \left(g^{1/2} \frac{4}{R_o^{\ 2}} \right) \right] \frac{\partial}{\partial\beta}$$

$$= \frac{4}{R_o^{\ 2}} \left\{ \frac{1}{\sin^2\beta} \left(\frac{\partial^2}{\partial\alpha^2} + \frac{\partial^2}{\partial\gamma^2} - 2 \cos\beta \frac{\partial^2}{\partial\alpha \, \partial\gamma} \right) + \right.$$

$$\left. + \frac{\partial^2}{\partial\beta^2} + \cot\beta \frac{\partial}{\partial\beta} \right\} \tag{E.41}$$

and

$$X_3(+) = \lambda_3^{\ \mu}(+) \, \partial_\mu = \frac{2}{R_o} \frac{\partial}{\partial\alpha} \tag{E.42+}$$

$$X_3(-) = \lambda_3^{\ \mu}(-) \, \partial_\mu = \frac{2}{R_o} \frac{\partial}{\partial\gamma} \quad . \tag{E.42-}$$

The Eulerian basis has been used in the text in equations (11.9) - (11.11) of Chapter 4 to discuss the rotation of a rigid body. An alternative way of obtaining these equations is suggested in problem 16 of Chapter 4.

(2) Stereographic Coordinate System

In Chapter 5 use was made of the stereographic coordinate system, which may be defined as follows

$$U = \frac{R_o + \frac{i}{2}\underset{\sim}{\sigma}\,\underset{\sim}{r}}{R_o - \frac{i}{2}\underset{\sim}{\sigma}\,\underset{\sim}{r}} \tag{E.45}$$

where the three ratios $\underset{\sim}{r}/R_o$ fix U. In this case one finds that the metric tensor is diagonal

$$g_{\alpha\beta} = G^2\ \delta_{\alpha\beta} \tag{E.46}$$

where

$$G^{-1} = 1 + \frac{1}{4}\ \frac{r^2}{R_o^{\ 2}} \quad . \tag{E.46a}$$

The invariant volume element in the group space is

$$d\tau = \sqrt{g}\ d^3a \quad .$$

In this case

$$d\tau = G^3\ d^3r \tag{E.47}$$

as remarked in Chapter 5 in connection with (10.8).

f) Irreducible Representations, Orthogonality Relations, and Green's Function

The irreducible representations satisfy the following equations:

$$X^2 D^j_{mm'}(\alpha\beta\gamma) = 4j(j+1)\frac{1}{R_o^{\,2}} D^j_{mm'}(\alpha\beta\gamma)$$

$$X_3(+) D^j_{mm'}(\alpha\beta\gamma) = \frac{2m}{R_o} D^j_{mm'}(\alpha\beta\gamma)$$

$$X_3(-) D^j_{mm'}(\alpha\beta\gamma) = \frac{2m'}{R_o} D^j_{mm'}(\alpha\beta\gamma)$$

We shall write these relations as differential equations in the following form

$$X^2 \mathcal{D}^j_{mn}(a) = \frac{4j(j+1)}{R_o^{\,2}} \mathcal{D}^j_{mn}(a)$$

$$X_3(+) \mathcal{D}^j_{mn}(a) = \frac{2m}{R_o} \mathcal{D}^j_{mn}(a)$$

$$X_3(-) \mathcal{D}^j_{mn}(a) = \frac{2n}{R_o} \mathcal{D}^j_{mn}(a) \qquad \text{(E.48)}$$

where the differential operators and the functions are supposed to be written in an arbitrary coordinate system.

The eigensolutions satisfy the orthogonality relation

$$\int \mathcal{D}^j_{mn}(a)^* \mathcal{D}^{j'}_{m'n'}(a)\sqrt{g}\, d^3a = \delta^{jj'}\delta_{mm'}\delta_{nn'}\frac{V}{d_j} \qquad \text{(E.49)}$$

(E.49) is the generalization of (D.16) to a continuous group. Here the group volume V replaces the number of elements g and

$$d_j = 2j + 1 \tag{E.49a}$$

$$V = \int \sqrt{g}\, d^3a \quad . \tag{E.49b}$$

V is the invariant volume of the group space and one may check by (D.47) that

$$V = 2\pi^2 R_o{}^3 \quad . \tag{E.50}$$

[The group space is a space of constant curvature; its radius of curvature is R_o and (E.50) is the usual formula for the volume.]

The Green's function is

$$G(a,a') = \sum_{jmn} \mathcal{D}^j_{mn}(a)\, \mathcal{D}^j_{mn}(a')^* \quad . \tag{E.51}$$

Consider

$$G_s(a,a') = \sum_{jmn} s^{2j}\, \mathcal{D}^j_{mn}(a)\, \mathcal{D}^j_{mn}(a')^* \quad . \tag{E.52}$$

Notice that

$$\sum_{m,n} \mathcal{D}^j_{mn}(a)\, \mathcal{D}^j_{mn}(a')^* = \sum \mathcal{D}^j_{mn}(a)\, \mathcal{D}^j_{nm}(a')^{-1} = \sum \mathcal{D}^j_{mm}(RR'^{-1})$$

since $\mathcal{D}^j_{mn}$ is a unitary representation where R and R' are the group elements corresponding to the coordinates a and a'. But

$$\sum_m \mathcal{D}^j_{mm}(RR'^{-1}) = \chi^j(RR'^{-1}) = \frac{\sin(j + 1/2)\omega}{\sin \frac{1}{2}\omega}$$

where χ^j is the trace or the character of the jth irreducible representation and ω is the magnitude of the rotation RR'^{-1}. (See problem 18, Chapter 4.)

Therefore we have the result:

$$G_s(a,a') = \sum s^{2j} \frac{\sin(j + 1/2)\omega}{\sin \frac{1}{2}\omega}$$

$$= [1 - 2s \cos \tfrac{1}{2}\omega + s^2]^{-1} \tag{E.53}$$

and

$$G(a,a') = \frac{1}{4 \sin^2 \frac{1}{4}\omega} \quad . \tag{E.54}$$

Let us also express this result in stereographic coordinates.

If U and U' are the two-dimensional representations of the rotations R and R', then

$$\mathrm{Tr}\ U'\ U^+ = \cos \tfrac{1}{2}\omega \tag{E.55}$$

by definition of ω. The preceding equation may also be expressed in the following form:

$$\mathrm{Tr}\ (U-U')(U-U')^{+} = 4\ \sin^2 \frac{1}{4}\ \omega \qquad \text{(E.56)}$$

and

$$G(r)\ G(R-R')^2\ G(r') = 4\ \sin^2 \frac{1}{4}\ \omega \qquad \text{(E.57)}$$

where $R = \underset{\sim}{r}\ \underset{\sim}{\sigma}$ and $\underset{\sim}{r}$ is now an abbreviation for the three stereographic coordinates. Therefore we have the final result

$$\sum \mathcal{D}^j_{mn}(r)\ \mathcal{D}^j_{mn}(r')^* = [G(r)\ G(\underset{\sim}{r}-\underset{\sim}{r}')^2\ G(r')\ G']^{-1} \qquad \text{(E.58)}$$

when the result is expressed in stereographic coordinates. This result has been used in Chapter 5, equation (10.9).

---

* This well known relation may be shown as follows:

$$\Gamma^{\alpha}_{\ \mu\alpha} = \frac{1}{2}\ g^{\alpha\sigma}\ \partial_\mu\ g_{\alpha\sigma} = \frac{1}{2}\ \mathrm{Tr}\ \underset{\sim}{g}^{-1}\ \partial_\mu\ \underset{\sim}{g}$$

where $\underset{\sim}{g}$ is the matrix $(g_{\alpha\beta})$. Let

$$\underset{\sim}{g} = e^{\underset{\sim}{\phi}} \quad .$$

Then

$$\det \underset{\sim}{g} = e^{\mathrm{Tr}\, \underset{\sim}{\phi}}$$

and

$$\ell n \det \underset{\sim}{g} = \mathrm{Tr}\, \underset{\sim}{\phi} = \mathrm{Tr}\, \ell n\, \underset{\sim}{g} \quad .$$

Therefore

$$\begin{aligned} \delta\, \ell n \det \underset{\sim}{g} &= \delta\, \mathrm{Tr}\, \ell n\, \underset{\sim}{g} \\ &= \mathrm{Tr}\, \delta\, \ell n\, \underset{\sim}{g} \\ &= \mathrm{Tr}\, \underset{\sim}{g}^{-1}\, \delta \underset{\sim}{g} \quad . \end{aligned}$$

It follows that

$$\partial_\mu\, \ell n \det \underset{\sim}{g} = \mathrm{Tr}\, \underset{\sim}{g}^{-1}\, \partial_\mu\, \underset{\sim}{g} \quad .$$

Finally

$$\begin{aligned} \Gamma^\alpha{}_{\mu\alpha} &= \frac{1}{2}\, \partial_\mu\, \ell n \det g \\ &= \frac{1}{\sqrt{g}}\, \partial_\mu \sqrt{g} \end{aligned}$$

where

$$g = \det \underset{\sim}{g} \quad .$$

INDEX

MODERN PHYSICS MONOGRAPH SERIES

FELIX M. H. VILLARS, Editor

Robert T. Schumacher
Introduction to Magnetic Resonance
Principles and Applications

This book, the first to discuss magnetic resonance at the undergraduate level, was developed as a supplement for courses in atomic, nuclear, or solid state physics, and as a reference for laboratory courses. It emphasizes semiclassical explanations of the characteristic phenomena of magnetic resonance, and illustrates the phenomena with specific examples taken from atomic, nuclear, and solid state physics, and chemistry. The juxtaposition of the characteristic features of magnetic resonance and their diverse applications relates theory to practice. Some of the problems following each chapter emphasize quantitative calculations, while others extend the theory to more advanced levels.

The wide range of subjects treated and the references to further work in each area make the book suitable to a broad audience. "Introduction to Magnetic Resonance: Principles and Applications" also provides an excellent introduction to standard graduate level texts.

1970, xiv, 220 pp., illus.
hardbound [ISBN: 0-805-38504-5]
paperbound [ISBN: 0-805-38505-3]

In Preparation:

HELLMUT J. JURETSCHKE
Crystal Physics